DARKNESS
AT
★ NIGHT ★

EDWARD HARRISON

DARKNESS AT NIGHT

★ A Riddle of the Universe ★

HARVARD UNIVERSITY PRESS

CAMBRIDGE, MASSACHUSETTS

LONDON, ENGLAND

1987

★

Library of Congress Cataloging-in-Publication Data

Harrison, Edward Robert.
Darkness at night.

Bibliography: p.
Includes index.
1. Astronomy—History.   2. Cosmology—History.
3. Astrophysics—History.   I. Title.
QB28.H37   1987   520′.9   86-32701
ISBN 0-674-19270-2 (alk. paper)

Designed by Gwen Frankfeldt

★

Preface

MY INTEREST in the tantalizing riddle of the dark night sky began many years ago. I puzzled over the problem of why the universe is not filled with light, and too soon thought I had it solved. But I was not to escape from the coils of this very old riddle so easily. Sometimes for hours, sometimes for days, I returned, lured by its power and subtlety. Each time new facets sprang into view. Little by little, I also explored the history of the riddle and discovered a wealth of literature on the subject, written over the centuries by scientists, philosophers, and poets.

I am grateful to many colleagues and correspondents—far too many to acknowledge by name—for their ideas and suggestions. I must limit my thanks here to those who contributed directly to the production of this work: Marie Litterer for her illustrations; Debbie van Dam and Debbie Schneider for assistance in translating works by Jean-Philippe Loys de Chéseaux and Heinrich Wilhelm Olbers; Virginia Trimble for her perceptive comments; Linda Arny, Alena Chadwick, Eric Esau, Marjorie Karlson, and Barbara Morgan of the University of Massachusetts Library, and Eleanor Brown of the Frost Library at Amherst College, for their advice and help; and P. A. Hingley of the Royal Astronomical Society (London) for assistance in investigating historical sources.

To my wife, Photini, for her unflagging support, I dedicate this book.

★

Contents

CONTENTS

★

Deep into the darkness peering, long I stood
there, wondering, fearing,
Doubting, dreaming dreams no mortal ever
dared to dream before.

Edgar Allan Poe

Prologue

S TARS gleam overhead on clear moonless nights, and darkness covers the valleys and hills. Why is the sky dark at night? The answer to this old and celebrated riddle seems deceptively simple: The Sun has set and now shines on the other side of the Earth. But suppose we were space-travelers and far from any star. Out in the depths of space the heavens would be dark, even darker than the sky seen from the Earth on cloudless, moonless nights. The riddle becomes: Why are the heavens not filled with light? Why is the universe plunged into darkness?

Astronomers have long pondered on the dark night-sky riddle and proposed many interesting answers. The search for the solution stretches over more than four hundred years; it explores immense spans of space and time, the nature of light, the structure of the universe, and other intriguing subjects. Misleading trails of inquiry and strange discoveries abound in the quest for the solution to the riddle of cosmic darkness.

Why Is the Sky Dark at Night?

How quaint the ways of paradox—
At common sense she gaily mocks.

William Gilbert, *Pirates of Penzance*

★

THE SKY at night is studded with twinkling stars. A few gleam brightly, for they are comparatively near or intrinsically luminous. Many only glimmer, for they are far away or intrinsically faint. Through telescopes—even field glasses—the scattered stars multiply, and we see myriads stretching away in all directions to immense distances.

The total number of stars seen with the unaided eye from all places on the Earth's surface is about 6,000 under the best "seeing" conditions. From any one place, the number drops to around 2,000—less than half, because we see at most only half the sky, and also we fail to see the faintest stars near the horizon. With field glasses, the number of stars observed from one place increases to 50,000 and even more, and rises to about 300,000 with a two-inch telescope.

Most stars belong to families of two or more members orbiting one another. Some are in loose clusters of hundreds of very young stars, such as Orion and the Pleiades, and we see them festooned with luminous wisps of gas. Many belong to less youthful clusters, such as Praesepe in the constellation of Cancer (Figure 1.1). Others belong to far-flung globular clusters, each a dense throng of hundreds of thousands of old stars born when our Galaxy was young. Beyond the Galaxy, strewn throughout an immensity of space, we see multitudes of other galaxies alit with stars.

But despite their vast number, stars fail to cover the whole sky. They stand out as isolated points of twinkling light against a background of

1.1   The Praesepe (Beehive) cluster in Cancer. These stars were about 400 million years old when they emitted the light that has taken roughly 500 years to reach us on Earth. (Courtesy of Mount Wilson and Las Campanas Observatories. Mount Wilson Observatory photograph.)

almost total darkness. The most meaningful observation that can be made in astronomy happens also to be the simplest: The sky at night is dark; gaps of darkness separate the stars. Why is the sky dark at night? In a universe that perhaps stretches away without limit and contains an unlimited number of stars, why do we not see starlight at every point of the sky? When observing the dark gaps that isolate stars, what are we actually seeing?

The Line-of-Sight Argument

To understand why the riddle of cosmic darkness has intrigued astronomers for centuries, we must first suppose that the universe has no edge but stretches away farther and farther, without end. (Never mind at this stage that space may be curved.) This supposition seems not unreasonable, for we find it difficult to imagine the universe ending abruptly at a cosmic edge, with no space beyond. We must next suppose that in this endless universe the stars stretch away endlessly. (Never mind for the moment that the majority of them huddle together in galaxies.) Because stars are not geometric points but luminous bodies of finite size like the Sun, we realize after some thought that a line of sight pointing in any direction ultimately intercepts the surface of a star (Figure 1.2). Some readers may find this difficult to understand. But consider: if space is *endless,* and if stars stretch away *endlessly,* any straight line from Earth *must* eventuallly encounter the surface of a star.

A forest is a helpful analogy. When we stand inside a deep forest and look around, we see trees in every direction. Tree trunks overlap and fuse to form a continuous background: every line of sight intercepts a tree trunk. If the forest were endless, then no matter how the trees were scattered and how widely they were separated, they would always form a continuous background. Even when clustered into groups, which in turn are clustered into larger groups, they would always fuse together, leaving no gaps. Our argument applies also to a forest of stars: If the star-strewn universe is endless, then no matter how the stars are scattered and how widely separated, they will always form a continuous background. Even when clustered into groups, which in turn cluster into galaxies, they will always fuse together, geometrically occulting one another, leaving no gaps (Figure 1.3). (Later we consider an exception to this rule in the hierarchical universe.)

The line-of-sight argument leads to the conclusion that the whole sky should glare with intense starlight. If most stars are like the Sun, every point of the sky should blaze as bright as the disk of the Sun. Intense starlight should pour down from a fiery canopy, as though the Earth were inside a white-hot furnace, or had plunged into the photosphere of the Sun. Calculation shows that the whole sky should be 180,000 times brighter than the disk of the Sun.[1] In this inferno of intense heat, the Earth's atmosphere would vanish in minutes, its oceans boil away in hours, and the Earth itself evaporate in a few years. And yet, when we survey the heavens, we find the universe plunged in darkness.

Riddles and Paradoxes

In 1952 Hermann Bondi reawakened interest in the riddle of cosmic darkness with his widely read book *Cosmology.* He attributed the discovery of the riddle to Wilhelm Olbers, a German astronomer of the last century, and referred to the riddle as "Olbers's paradox."

Riddles can be puzzles, problems, or paradoxes. Generally they may

1.2 A line of sight in an endless universe. If the universe has no end, and if stars stretch away endlessly, then every line of sight eventually intercepts the surface of a star. Why, then, is the sky dark at night?

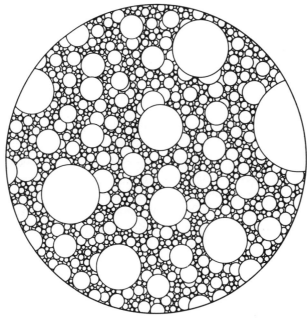

1.3 A bright, starlit universe in which stars cover every part of the sky. No dark gaps exist between the stars and the whole sky blazes with 180,000 times the light of the Sun.

be expressed as questions, as in the Sphinx's famous riddle to Oedipus, "What is it that walks on four legs in the morning, on two at noon, and on three in the evening?" (Oedipus solved the riddle by answering, "Man crawls as a child, stands erect in manhood, and leans on a staff in old age." In rage, the Sphinx threw herself over a cliff, and Oedipus, to his woe, became king of Thebes.) Often riddles may be statements having deep or dark meaning, as when Epimenides the Cretan said that all Cretans are liars, which, if true, made him a liar for telling the truth. Some riddles, such as this, lie deep at the foundations of formal logic.

All paradoxes are riddles, but not all riddles are paradoxes. Paradoxes consist of propositions contrary to known facts or received opinions. They startle us with contrast and contradiction. Thus, Zeno of Elea in the fifth century B.C. devised various arguments to prove the unreality of motion and change. The most famous of Zeno's arguments is the riddle of Achilles and the tortoise. The fleet-footed Achilles and the slow-moving tortoise hold a race, and the tortoise is allowed to start first. After the tortoise has traveled 100 units of distance, Achilles starts and soon covers this distance. In the meantime, the tortoise has crawled an extra unit of distance; and while Achilles runs this unit, the tortoise crawls an extra hundedth of a unit; and so on, with Achilles never able to reach and overtake the tortoise. How does Achilles win the race? This is a riddle; also a paradox, because plain for all to see in the phenomenal world is the fact that fast-moving objects overtake slow-moving objects. Zeno's celebrated paradoxes contradict common sense, but are by no means meaningless; at issue is whether continuous mathematical functions can represent our experience of movement.[2]

Life abounds with paradox, with paradox lost and paradox regained, yet "What is life?" is a riddle, not a paradox. The Victorian scientist Lord Kelvin, who figures prominently in our story about the riddle of cosmic darkness, said, "In science there are no paradoxes."[3] This down-to-earth realist believed that paradoxes exist in ourselves, not in the external world.

Two Interpretations

From the outset, we have a choice of two interpretations of the dark night sky: either, contrary to observation, the dark gaps are filled with stars that for puzzling reasons remain unseen, and we must account for the missing starlight; or, in accord with observation and contrary to expectation, the dark gaps are mostly empty of stars, and we must account for the missing stars. The "covered-sky" interpretation might be called the riddle of the missing starlight (Figure 1.4), and the "uncovered-sky" interpretation might be called the riddle of the missing stars (Figure 1.5). Both interpretations rank as riddles, but the first, because it assumes the opposite of what is observed, amounts to a paradox. The habit of referring to the riddle as Olbers's paradox favors the first (the covered-sky) interpretation, and we tend to overlook the possibility of the second (the uncovered-sky) interpretation.

Each interpretation defines a class of proposed solutions to the riddle of darkness. According to the first interpretation (which accepts the line-of-sight argument), rays of light emitted by the background of stars hurry toward us, and yet for puzzling reasons never reach the Earth. We will see that the number of stars needed to cover the whole sky is roughly 1 followed by 60 zeros, or one trillion trillion trillion trillion trillion (where one trillion is 1,000 billion, and one billion is 1,000 million). By comparison, the grains of sand in all deserts and on all beaches

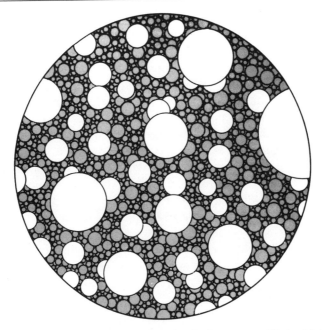

1.4 The "covered-sky" interpretation: The dark gaps are filled with invisible stars (gray areas). Where is the missing starlight?

of Earth number at most a trillion trillion. Each proposed solution to the riddle that accepts the covered-sky interpretation must explain why the starlight emitted by multitudes of stars fails to reach us. For example, Thomas Digges in the sixteenth century, and many astronomers who followed him, said the rays from the remotest stars are much too feeble to be detected by the eye. Loys de Chéseaux in the mid-eighteenth century and Olbers himself in the early nineteenth century said that distant stars are obscured from view by absorbing matter distributed throughout space. Hermann Bondi has said the rays from distant stars are redshifted into invisibility owing to the expansion of the universe. All proposed solutions in this class assume that stars cover the entire sky, and each solution offers an explanation for the missing starlight. Alas! when carefully examined, these and other solutions that seek to explain the missing starlight fail in the modern universe.

The second interpretation adopts the apparently simpleminded view that the dark gaps are indeed mostly empty. Of course, larger and better telescopes reveal more and fainter stars and also numerous extragalactic systems of stars, but however far we may eventually peer into the depths of space, according to this interpretation, we will always see stars immersed in pools of darkness. Each proposed solution to the riddle that accepts this interpretation attempts to explain why the observed dark gaps are mostly empty and not filled with stars. For example, Johannes Kepler in the early seventeenth century said we look out between the stars and see a dark wall enclosing the universe. Later in the same century, Otto von Guericke said the universe consists of a starry cosmos of finite size surrounded by an infinite void, and we look out between the stars and see the dark emptiness of space beyond the cosmos. John Herschel and Richard Proctor in the nineteenth century said we live conceivably in a hierarchical universe arranged in larger and larger stellar systems in such a way that distant stars, even though infinite in number, remain insufficient to cover the sky. Again alas! when carefully examined, these and many other solutions that seek to explain the missing stars also fail in the modern universe (Figure 1.6).

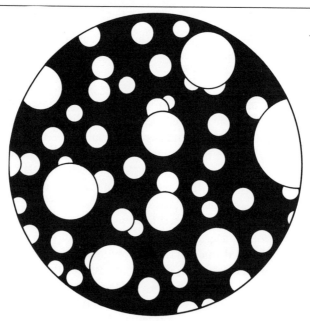

1.5   The "uncovered-sky" interpretation: The dark gaps are mostly empty of stars. Where are the missing stars?

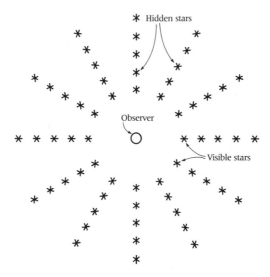

1.6   An improbable solution of why the sky is dark at night, proposed in jest by Edward Fournier d'Albe in 1907: An infinite number of invisible stars are aligned in rows behind the visible stars. Thus, the sky is dark in accordance with the uncovered-sky interpretation. (See entry 1 in the table "Proposed Solutions.")

Trails of Inquiry

Obviously, there must be a correct solution, and presumably it falls in one of these two classes. We shall find, despite its subtlety, that the riddle has a rather simple solution, which can be understood without a deep knowledge of science and mathematics. In fact, the first person to give a qualitatively correct solution was a nineteenth-century poet.

Does the star-filled universe extend to infinity? Does it contain obscuring matter that prevents us from seeing trillions and trillions of distant stars? Has the universe hierarchical structure? Has it flat or curved space? We must follow these and many other trails of inquiry to discover whether any leads to a satisfactory solution of the riddle of darkness. Pitfalls, mazes, dragons, and other hazards beset these trails and have trapped many an unwary explorer.

Darkness at Night divides into three parts. In the first part we trace the emergence of strands of thought in the ancient world that led to the origin of the riddle of darkness. We see scholarly divines in the Middle

Ages developing ideas about space and the universe that greatly influenced the course of science, and we see inspired astronomers in the sixteenth and seventeenth centuries formulating the riddle.

In the second part the Cartesian and Newtonian systems rise in the seventeenth century from the ashes of the Aristotelian, Epicurean, and Stoic systems of the ancient world, and the unfolding drama of science deepens and enriches the riddle, disclosing the possibility of more than one solution.

In the third part the strands of our story converge and interweave with the discovery of galaxies, the rise of the new astronomy, and the emergence of modern cosmology. A version of the correct solution to the riddle is perceived vaguely and qualitatively by Edgar Allan Poe in the nineteenth century and clearly and quantitatively by Lord Kelvin in the early twentieth century. Strange to say, the Poe and Kelvin versions are either ignored or soon forgotten. In the twentieth century the riddle is temporarily overshadowed by the struggle to understand the physical and mathematical complexities of an expanding and evolving universe of curved space. Knowledge of the universe grows, the cosmological models of yesteryear's fashion shows multiply, and observational discoveries reveal that we live in an expanding universe suffused with the afterglow of the big bang. Then Hermann Bondi and other promoters of the steady-state universe resurrect the riddle. For a time it seems that darkness at night offers startling proof of the expansion of the universe. "Go out at nighttime and note the darkness of the heavens; their darkness proves that the universe is expanding!" Unfortunately, this ingenious solution applies only in an expanding steady-state cosmological model—now controverted by the discovery of the afterglow of the big bang. The sudden realization that the universe in its present state lacks enough energy to create a bright starlit sky becomes of paramount importance. It overrides all other considerations and vanquishes almost all previous explanations. With fresh insight, we realize that the Poe–Kelvin solution strikes to the heart of the riddle of cosmic darkness.

PART ONE

★ THE ★
RIDDLE
BEGINS

Three Rival Systems

I do not hesitate to assert that I consider astronomy as the most important force in the development of science since its origin sometime around 500 B.C.

Otto Neugebauer, *The Exact Sciences in Antiquity*

THREE systems of natural philosophy dominated the Mediterranean world in the pre-Christian era. They emerged from the germinal thoughts of the Ionians and Pythagoreans and have shaped the history of Western culture and science. All three have particular interest for us in our pursuit of the origin and solution of the riddle of cosmic darkness.

The atomist system of endless space strewn with numberless worlds composed of atoms came first; followed by Aristotle's harmonious geometric system of celestial spheres; followed finally by the Stoic system of a starry cosmos surrounded by an infinite extracosmic void. The atomist system was adopted and elaborated by Epicurus and his followers; the Aristotelian system was adopted and developed by the philosophers at the Museum in Alexandria; and the Stoic system was adopted and admired by Roman intellectuals. Ingredients of all three systems survive in the modern world. We inherit from the atomists–Epicureans the corpuscular constitution of matter and the concept of endless space, from the Aristotelians the rhythms of an orderly natural world, and from the Stoics the picture of gyrating worlds in a cosmic void and the theory of swirling fluids. Entangled in these systems come religious, philosophical, and ethical legacies; but that is an entirely different story.

Atomists and Epicureans

Perhaps, like so many other ideas, the atomic theory of matter began with Pythagoras of Samos in the sixth century B.C. He and his followers, the Pythagoreans, held that all physical things are composed of geometric points having mathematical relations.

In the fifth century B.C. the atomic theme was pursued by Anaxagoras, the last of the Ionian natural philosophers from the Turkish coast of the Aegean. He lived and taught in Athens during its golden age and was a friend of the ruler Pericles, and also of Pericles's gifted consort, Aspasia, born in Miletus near Anaxagoras's hometown, Clazomenae. The Mind, declared Anaxagoras, holds sway over an immeasurable universe in which everything on Earth and in the heavens consists of an assortment of minute, seedlike components and obeys universal laws.

Leucippus transformed the Pythagorean points and Anaxagorean seeds into ultimate physical entities called atoms. ("Atom" means uncuttable.) Nothing is known about his life, and apart from fragmentary comments by later philosophers, almost nothing is known about his philosophy. He is overshadowed by Democritus of Abdera, who lived in Athens at the time of Socrates and further developed the atomic theory. Democritus declared that atoms are the smallest possible subdivisions of matter, and their combinations and mathematical relations account for the properties of all things of sensible size. Only atoms and the infinite void exist, all else is opinion of the mind and convention of the senses. Records state that Democritus ranked intellectually equal if not superior to Plato and that his works, now vanished, were as comprehensive as those of Aristotle.

Hanging on the coattails of the atomic theory came the concept of infinite space. Conceivably, concourses of atoms conjure up the prospect of endless repetition; beyond the horizon the universe appears much the same as close at hand; journeys, however vast, lend no surprise. Shapes may change and details vary, but the fundamentals in the cosmic design remain ever the same. The idea of endless repetition in the firmament eventually developed into the concept of cosmic uniformity that is now the foundation stone of modern cosmology and known as the *cosmological principle*. The atomists believed that the natural world extends without limit beyond the control of finite gods and is self-governing.

According to Plutarch in his essay "The tranquility of the mind," Alexander the Great wept when told by Anaxarchus of Abdera, an atomist, that there exists an infinity of worlds.[1] To his companions the weeping Alexander said, "Do you not think it a matter worthy of lamentation that when there is such a vast multitude of them, we have not yet conquered one?" The idea of an infinite universe teeming with a plurality of worlds has reverberated around the globe for twenty-five centuries, transforming science, philosophy, and religion.

Epicurus of Samos established in Athens in the late fourth and early third centuries B.C. a freethinking school that embraced much of the atomist philosophy (Figure 2.1). He rejected the gods as the controlling forces of the natural world, invoked physical causes whenever possible, and taught that sense perceptions form the basis of all knowledge. Epicurus also outfitted the atomist world of natural laws with a comprehensive theory of ethics, claiming the highest good to be pleasure taken wisely and in moderation, leading to a life of maximum freedom from physical and mental pain.[2] The atomist system, taught in this practical form—unfortunately easily misunderstood as a hedonistic doctrine—gained the acceptance of multitudes throughout the known world. Educated Romans, skeptical of the mythological religions, turned to the teachings of Epicurus. For more than seven centuries, three before and four after Christ, Epicureanism flourished. The Epicureans revered Epicurus as a savior from superstition and ignorance, and displayed his likeness in pictures and statues in their homes. But others—first the Platonists, then the Stoics, and finally the Christians—reviled him as an atheist and attacked his doctrines with bitter hostility.

The Epicurean Roman poet Titus Lucretius Carus, born soon after 100 B.C., had probably already died when his epic poem *De Rerum Natura* (*On the Nature of Things*) was made public about 55 B.C. Cyril Bailey writes that the poem reveals Lucretius's "fierce hatred of conventional superstitions and a yearning for intellectual liberty in the presence of nature." The poem gives a vivid account of the atomist system. Eternal atoms move freely through an infinite void, colliding, combining, and forming the substantial fabrics of countless worlds. Suns brighten, shine for eons, fade, and dissolve back into the atomic ferment. Atoms combine, worlds form, mountains thrust up above seas, life originates, organisms evolve, intelligent creatures emerge, societies develop, civilizations rise, and then ultimately each world dissolves back into the welter from whence it came. Only atoms and the void endure forever unchanged.[3]

The soul consists of atoms, wrote Lucretius, and the gods, if they exist, also consist of atoms. Occasionally atoms swerve unpredictably when encountering one another, and such indeterminate events account for chance and acts of free will.

2.1 Epicurus (341–270 B.C.), founder of the Epicurean-atomist school, born in Samos, but lived in Athens most of his life.

Aristotle Creates Order

Whatever his views concerning endless space, we know that Pythagoras said the Earth is a sphere, and the celestial bodies move in perfect circular orbits around a central cosmic fire. He called the universe a cosmos—a harmonious whole—and sought in the tumult of the world to find an inner numerical harmony. To this end he developed the mathematics of Mesopotamia and Egypt to the level later inherited by Euclid.

From the Pythagorean celestial orbits and perfect motions Plato and scholars of the Academy in Athens constructed a cosmic geometry of celestial spheres having the Earth at the center and the whole enclosed in an outermost sphere of affixed stars. This cosmic scheme, conceived in the fourth century B.C., formed the basis of the second great system of natural philosophy of the ancient world (Figure 2.2).

Aristotle, at first a student at the Academy and then tutor to the youthful Alexander of Macedonia before Alexander began conquering lands to the east, established in Athens his own institute of learning called the Lyceum (Figure 2.3). He developed the Earth-centered many-sphere system of the Academy into an orderly physical-ethereal finite universe ruled by the eternal ideas.[4]

Aristotle taught that the celestial spheres surrounding the Earth consist of a single incorruptible element, ether, having imperishable forms and perfect circular motions. These spheres support in ascending order the "wandering stars" (the literal meaning of *planets*) comprising the Moon, Mercury, Venus, Sun, Mars, Jupiter, and Saturn, which rotate about inclined axes at various rates. Ethereal light fills the vault of heaven, and beyond the outermost sphere of affixed stars exists nothing, "neither place nor void nor time."

The Earth and its sublunar regions, said Aristotle, consist of the four corruptible elements fire, air, water, and earth having perishable forms and jerky imperfect motions. Fire, by virtue of levity, seeks the heavens; earth, by virtue of gravity, seeks the center of the world; air and water float between the two extremes.

Astronomers, notably Hipparchus with his observatory at Rhodes in the second century B.C. and Claudius Ptolemy at the Museum in Alexandria in the second century A.D., added geometric elaborations to the Aristotelian system. Neoplatonists of Alexandria contributed angelic ornamentations and even constructed an analogous system of angelic spheres with God at the center instead of the Earth. Arab scholars studied and preserved the Aristotelian system through the Middle Ages, and

2.2 The spherical Earth occupied the center of the Aristotelian-Ptolemaic system, surrounded by the celestial spheres. The sphere of fixed stars formed a wall enclosing the universe, with nothing, neither space nor time, outside. This illustration from Edward Sherburne's *The Sphere* (1675) shows the addition of the *primum mobile*, an outermost sphere moved by God, introduced by medieval Arab astronomers.

in the high and late Middle Ages Aristotle's celestial spheres formed the basis of Islamic, Judaic, and Christian cosmology, but with important modifications, as we shall see.

The Starry Cosmos of the Stoics

The third great system was the popular Stoic world view founded by Zeno of Citium, who lectured in a roofed colonnade called a stoa (Figure 2.4). From its beginning the Stoic philosophy appealed to all classes,

2.3   Aristotle (384–322 B.C.), born in Stagira, northern Greece, founder of the Lyceum in Athens.

2.4 Zeno of Citium (c.334–c.262 B.C.), founder of the Stoic school, born in Cyprus and lived in Athens. Known as Zeno the Phoenician and Zeno the Stoic.

2.5 What happens to a spear thrown across the edge of the universe? This riddle, posed by Archytas, drew attention to the problem of space terminating abruptly at a boundary. (E. R. Harrison, *Cosmology,* courtesy of Cambridge University Press.)

from slaves to aristocrats. Have fortitude in the face of adversity for Fate rules the world! Weep not for thou art strong. The gods exist as spirits of nature, or within ourselves, and the divine spirit throbs on Earth and in the heavens, swelling and subsiding from age to age, from cycle to cycle on the Wheel of Time. Gaze on it all but be not amazed for the soul has witnessed it many times before.

The Stoic system exalted the ethical principles of duty and justice, as exemplified in the writings of Seneca and Marcus Aurelius, and served as everyone's philosophy, religion, ethics, and science. Its concepts, values, and codes of honor have infiltrated Western culture more deeply than we commonly recognize.[5] The Stoic world view served as the ancient counterpart of the modern popular scientific world view.

Archytas of Tarentum, a Pythagorean and friend of Plato, posed a riddle: "What happens when a spear is thrown across the edge of the universe? Does it rebound or vanish from this world?" Archytas's riddle of the cosmic edge, showing how illogical is the belief that whatever bounds space is itself unbounded, was a recurrent theme in the history of science for the next two thousand years (Figure 2.5).[6] The Stoics fully

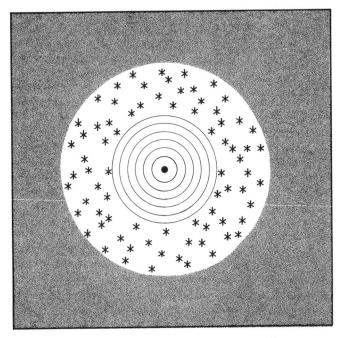

2.6   In response to the riddle posed by Archytas, the Stoics proposed an island cosmos of stars surrounded not by an edge but a void of infinite extent. The Stoic system persisted in modified forms until the twentieth century.

agreed that space has no edge and rejected the Aristotelian outer boundary of the universe. They proposed instead a system consisting of a star-filled cosmos surrounded by a starless extracosmic void extending to infinity. In the ancient world and in the late Middle Ages the Stoic system was more or less the Aristotelian system of celestial spheres stripped of its outer boundary (Figure 2.6).

The Stoic astronomical scheme endured in various guises for more than two millennia until the first quarter of this century, when the existence of galaxies beyond the Milky Way was established beyond dispute. From the seventeenth and possibly the sixteenth century it offered to many astronomers what seemed a natural explanation of a dark night sky: we look out between the stars, beyond the cosmos, and see the darkness of an infinite extracosmic void. The Stoic system formed the framework of nineteenth-century cosmology during the rise of the new astronomy. In the early decades of this century, astronomers such as the young American Harlow Shapley looked up at the night sky and saw beyond the Stoic-galactic edge of the Milky Way an interminable sea of empty darkness.

Large optical and radio telescopes reaching out beyond the Galaxy and Local Group of galaxies, far beyond the Local Supercluster of galaxies, into the depths of space, have not found an edge to the star-filled universe, beyond which lies a void. We know at last that we do not live in the Stoic cosmos.

The Cosmic-Edge Riddle

Historical records contain numerous references to Archytas's cosmic-edge riddle, and without doubt it opened a new world of inquiry. The Pythagorean's riddle is of utmost importance in our story, for it led ultimately to the dark night-sky riddle. Some extracts show how its fascination remained undiminished over the centuries.

Simplicius, among the last of the Neoplatonists in the sixth century, quoted Archytas in his commentary on Aristotle's *Physics:* "If I am at the extremity of the heaven of the fixed stars, can I stretch outwards my hand or staff? It is absurd to suppose that I could not; and if I can, what is outside must be either body or space. We may then in the same way get to the outside of that again, and so on; and if there is always a new place to which the staff may be held out, this clearly involves extension without limit." This fragment remained untranslated into Latin until the sixteenth century.[7]

Lucretius wrote in *De Rerum Natura,* "Learn, therefore, that *the universe is not bounded in any direction.* If it were, it would have a limit somewhere. But clearly a thing cannot have a limit unless there is something outside to limit it . . . It makes no odds in which part of it you may take your stand: whatever spot anyone may occupy, the universe stretches away from him just the same in all directions without limit." He recalled the argument by Archytas: "Suppose for a moment that the whole of space were bounded and that someone made his way to its uttermost boundary and threw a flying dart. Do you choose to suppose that the missile, hurled with might and main, would speed along the course on which it was aimed? Or do you think something would block the way and stop it? You must assume one alternative or the other. But neither of them leaves you a loophole. Both force you to admit that the universe continues without end. Whether there is some obstacle lying on the boundary line that prevents the dart from going farther on its course or whether it flies on beyond, it cannot in fact have started from the boundary. With this argument I will pursue you. Wherever you may place the ultimate limit of things, I will ask you: 'Well then, what does happen to the dart?'"[8]

In the Middle Ages, Archytas's cosmic-edge riddle was known from the commentary by Simplicius on Aristotle's *De Caelo* (*The Heavens*): "Stoics, however, thinking that there is a vacuum beyond the sky, prove it by this kind of assumption: let it be assumed that someone standing motionless at the extremity of the world extends his hand upward. If his hand does extend, they take it that there is something beyond the sky to which the hand extends. If the hand cannot extend, then something exists outside that prevents its extension. But if he then stands at the extremity of this obstacle that prevents the extension and then extends his hand, the same question as before must be asked."[9] Many divines, including Thomas Aquinas, in the high and late Middle Ages referred to this comment by Simplicius.

Burchio, a stubborn Aristotelian character in Giordano Bruno's *Infinite Universe and Worlds* (1584), argued, "If a person would stretch out his hand beyond the convex sphere of heaven, the hand would occupy no position in space, nor any space, and in consequence would not exist." But Philotheo (Bruno himself) replied that space inside and outside must be continuous, and added, "Thus, let the surface be what it will, I must always put the question: what is beyond?"[10] We shall see that Bruno in this and other works pioneered a centerless and edgeless universe of countless inhabited worlds.

John Locke repeated the same argument in *An Essay Concerning Human Understanding* (1690): "If Body be not supposed infinite, which I think no one will affirm, I would ask, Whether, if God placed a Man at the extremity of corporeal Beings, he could not stretch his Hand beyond his Body? If he could, then he would put his Arm where there was before Space without Body; and if there he spread his Fingers, there would still be Space between them without Body: If he could not stretch out his Hand, it must be because of some external hindrance . . . And then I ask, Whether that which hinders his Hand from moving outwards be Substance or Accident, Something or Nothing? . . . For I would fain meet with that thinking Man that can, in his Thoughts, set any bounds to Space, more than he can to Duration; or by thinking, hope to arrive at the end of either."[11]

Geometers thought that Archytas, by demonstrating the continuity of space, had proved that space is infinite. They supposed that unbounded space must extend in all directions to infinity and obey Euclidean rules of geometry. But the demonstration that space cannot terminate at an edge proved only that space is unbounded. Since the middle of the nineteenth century we have known that unbounded or edgeless space need not be infinite and need not obey Euclidean rules of geometry. An example of an alternative possibility is a finite but unbounded three-dimensional space analogous to the finite but unbounded two-dimensional surface of a sphere. A Spherelander—a two-dimensional creature living on the surface of a sphere—occupies an edgeless world that nonetheless is finite.[12]

Celestial Light

And lest thou lift up thine eyes unto heaven, and when thou
seest the sun, and the moon, and the stars, *even* all the host of
heaven, shouldest be driven to worship them, and serve them.

Deuteronomy 4:19

BARBARIAN victories and the spread of Christianity accompanied the
decline of the Roman Empire. While bureaucratic Byzantium in Eastern
Europe hoarded much of ancient knowledge, it added little, and be-
grudged sharing what it had with barbarians. The pursuit of knowledge
languished and intellectual gloom descended on the Mediterranean
world. Movements of utmost social importance were nonetheless afoot;
slavery retreated slowly before the spread of Christianity, Benedictine
monks established monastery schools in the sixth century, and a Western European cultural unity emerged in the succeeding centuries.

From Baghdad, Cairo, Cordoba, and other centers of learning in the
flourishing Islamic Empire came disturbing whispers of knowledge that
awoke Western Europe. A technological revolution began and a society
emerged in the twelfth and thirteenth centuries unlike any in the ancient world.[1] Mills harnessed the power of rivers, tides, and winds, ca-
thedrals aspired to the nine skies, and townfolk and countryfolk became
artisans skilled in hundreds of crafts and trades.

Arabs, then Jews, and then Christians, in their philosophical and
theological studies, adopted the Aristotelian system of concentric celestial spheres. The Arabs created the primum mobile, an outer sphere that
transmitted motion to all other spheres and was itself driven by divine
will. Anselm in the eleventh century, archbishop of Canterbury, introduced the empyrean, an outermost sphere of purest fire where God
dwelt (Figures 3.1–3.4).

The high Middle Ages became an age of scribes translating Arab and
Greek manuscripts. The new knowledge exceeded the scope of the

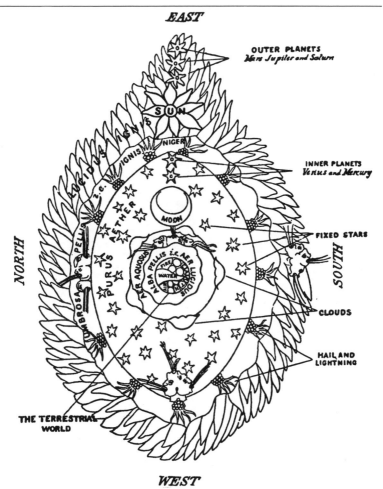

3.1 The universe according to Hildegard of Bingen in Germany. In her writings we see how the influx of knowledge in the twelfth century modified the medieval Christian picture of the universe, a picture that developed and climaxed in Dante's *Divine Comedy*. (Reproduced from the Wiesbaden Codex B as figure 2 in Charles Singer, "The scientific views and visions of Saint Hildegard.")

monastery and cathedral schools, and communities of learned scholars founded the universities.[2] Students flocked to these centers of learning, and the translated works of Aristotle, Euclid, Ptolemy, Galen, and other sages of the ancient world revealed vistas of knowledge that exalted the power of the human mind.

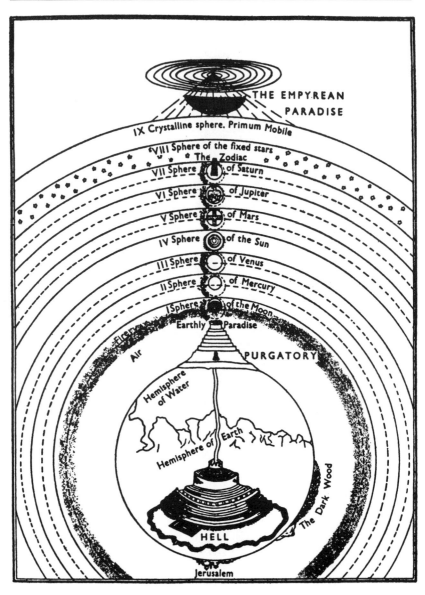

3.2 Dante's system as usually represented. (Reproduced from figure 4 in Charles Singer, "The scientific views and visions of Saint Hildegard.")

3.3 The "Empyrean" by Gustave Doré (1832–1883), showing Dante and Beatrice gazing upon the theocentric world of angelic spheres from the rim of the geocentric world of celestial spheres.

3.4 "The Creation of the World" from the four-volume encyclopedia on human knowledge (1617–1619) by Robert Fludd, a physician of early seventeenth-century London. Adam and Eve stand in the center of the Earthly paradise, surrounded by the celestial spheres and the sphere of fixed stars, beyond which lies Anselm's sphere of purest fire, known to poets after the time of Milton as the empyrean.

Omnipotent and Ubiquitous

But ecclesiastical authorities grew alarmed. The wholesale adoption of ancient beliefs about the nature of the universe threatened to transform Christian doctrine beyond recognition. Granted that the Earth rests at the center of the universe, and granted that the heavens consist of rotating etheric spheres, but to go further and assert that even God, if he so willed, could not move the Earth, or could not create other worlds than the Earth, as taught by professors in the schools of art, controverted the tenets of sacred doctrine. In 1277 the bishop of Paris, Etienne Tempier, issued the famous 219 condemnations of all professors who dared to place limits on the power and scope of the supreme being.

Pierre Duhem, French physicist and historian of science in the late nineteenth and early twentieth centuries, showed that the condemnations of 1277 stand as a landmark in the history of cosmology.[3] They inspired scholarly divines in the high and late Middle Ages to seek a more ample system to accommodate the works of a supreme being of greater power and extent. One might say that the history of medieval cosmology from start to finish consists of the struggle to match an abstract plentitudinous supreme being with Hebraic and Aristotelian scriptures.[4] By proclaiming Aristotle's fallibility and by undermining his already besieged finite anthropocentric system, the bishop's condemnations alerted scholars and divines to the need for a system more compatible with the idea of an omnipotent and omniscient supreme being.

By the fourteenth century, Thomas Bradwardine of Merton College at Oxford, who later became archbishop of Canterbury, could write, "God is that whose power is not numbered and whose being is not enclosed." He echoed earlier medieval scholars, who in turn echoed Empedocles of the fifth century B.C., when he wrote, "God is an infinite sphere whose center is everywhere and circumference nowhere." Bradwardine took the important step of extending the empyrean into an infinite extramundane void. Beyond the sphere of stars stretched a mysterious and limitless realm pervaded by spirit. He quite literally transformed the bounded Aristotelian system into an unbounded Stoic system.[5] From the fourteenth century onward, a mysterious extramundane void existed beyond the sphere of fixed stars, and with this major modification the Aristotelian universe bore certain resemblances to the Stoic system.

A century later Nicholas of Cusa (1401–1464)—a German cardinal and statesman—laid the foundation of post-medieval cosmology in his work *On Learned Ignorance*.[6] He summoned up the full potentiality of an omnipotent being to create the actuality of an unbounded universe. Because God is limitless and everywhere, he said, the universe must be edgeless and centerless. From Empedocles's and Bradwardine's principle comes the corollary, "The universe has its center everywhere and its circumference nowhere." Only with impiety dare we attribute less to the handiwork of God. The discovery in 1427 of the startling *De Rerum*

Natura by Lucretius and the publication of the imaginative *Learned Ignorance* in 1440 by Nicholas of Cusa opened the way for the infinite
universes of Giordano Bruno in the sixteenth century and of René Descartes and Isaac Newton in the seventeenth century.

By comparison with the bold steps into a centerless universe taken by
the scholarly divines, Nicholas Copernicus's revival of the old but still
unforgotten theory of a Sun-centered universe seems quite modest, and
at first glance hardly revolutionary in scope. The Earth-centered system
of Aristotle and Ptolemy was still the dominant cosmological model of
the early fifteenth century. Copernicus was inspired while a university
student by the thought that the Sun-centered system, proposed in
the third century B.C. by Aristarchus of Alexandria, might offer certain
computational advantages and reveal more harmony than an Earth-
centered system. For years he labored at calculating heliocentric planetary orbits, and his final work, *De Revolutionibus Orbium* (*The Revolutions
of the Celestial Orbs*, published in 1543, the year of his death at age seventy, greatly impressed astronomers and mathematicians (Figure 3.5).
Yet for several decades, like an unexploded bomb, it made little impact
on other sections of society.

Positioning the Sun at the center of the universe demoted not only
the Earth but also the sphere of fixed stars.[7] The starry sphere, now
stationary, no longer acted as an essential component in the celestial
machinery, but began to dissolve and disperse into the extracosmic void.
The important step of dispersing the stars was taken by Thomas Digges,
the foremost astronomer and one of the ablest mathematicians in England in the second half of the sixteenth century.

Digges Enlarges

When the popular astronomical guide *Prognostication Everlastinge* by
Leonard Digges was republished in 1576, Thomas Digges revised his
father's book and appended to it a short work entitled, "A Perfit Description." This work (reproduced here as Appendix 1) bears the full title, "A
Perfit Description of the Caelestiall Orbes, according to the most aunciente doctrine of the Pythagoreans, lately revived by Copernicus and
by Geometricall Demonstrations approved." In this work Thomas
Digges translated several significant passages of Copernicus's *De Revolutionibus Orbium* and included his own heliocentric diagram of the universe (Figure 3.6). He championed the heliocentric system, and by writing in English, rather than Latin, he communicated the new ideas in
astronomy to a wide audience.[8]

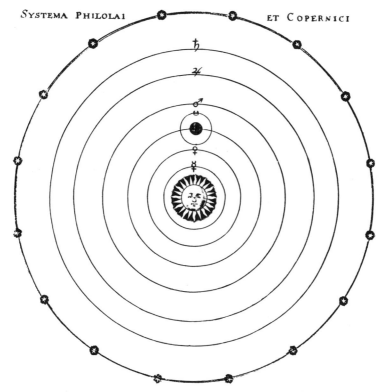

3.5   The heliocentric Copernican system of celestial spheres, from *Institutin Astronomica* by Pierre Gassendi (1592–1655), an innovative French thinker who revived and promoted the Epicurean-atomist system.

Copernicus had said little or nothing about what lay beyond the sphere of fixed stars. Digges's original contribution to cosmology consisted of dismantling the starry sphere and scattering the stars throughout endless space. He modestly made no claims of his own, and many of his readers probably assumed that this novel idea had originated with Copernicus. In Digges's diagram, the following four lines replace the sphere of fixed stars:

> This orbe of starres fixed infinitely up extendeth hit self in altitude spherically, and therefore
> Immovable the pallace of foelicitye garnished with perpetuall shininge glorious lightes innumerable
> Farr excellinge our sonne both in quantitye and qualitye the very court of coelestiall angelles

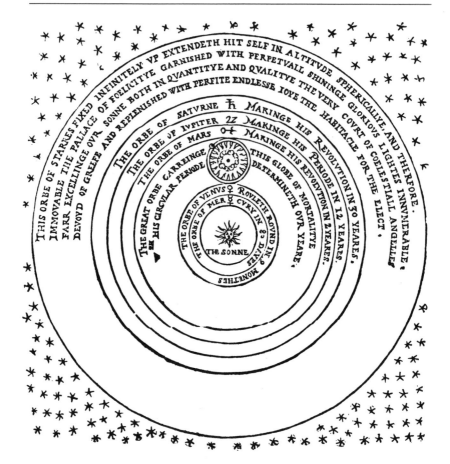

3.6   Thomas Digges's cosmographical diagram in the "Perfit description of the Caelistiall Orbes" (1576), which shows the stellar sphere dismantled and the stars dispersed throughout infinite space.

> Devoyd of greefe and replenished with perfite endlesse joye the habitacle
>   for the elect

His diagram shows an endless space beyond the planetary orbits, and in this space, occupied by innumerable stars excelling the Sun in quantity and quality, lay the glorious court of God.

Digges wrote in the "Perfit Description," "We may easily consider what little portion of gods frame, our Elementare corruptible worlde is, but never sufficiently be able to admire the immensity of the Rest."

Especially of that fixed Orbe garnished with lightes innumerable and reachinge up in *Sphaericall altitude* without ende. Of which lightes Celestiall it is to bee thoughte that we onely beholde sutch as are in the inferioure partes of the same Orbe, and as they are hygher, so seeme they of lesse and lesser quantity, even tyll our sighte beinge not able farder to reache or conceyve, the greatest part rest by reason of their wonderfull distance invisible unto us.

By grafting endless space onto the Copernican system and scattering the stars throughout this endless space, Digges pioneered in sixteenth-century astronomy the idea of an unlimited universe filled with the mingling rays of countless stars. Furthermore, he was the first person to formulate the riddle of darkness. He realized that the invisibility of an unlimited number of distant stars called for some sort of comment, and responded with the remark that most stars cannot be seen because "the greatest part rest by reason of their wonderfull distance invisible unto us." What could be more natural, given the state of optical science in the sixteenth century, than the simple idea that the most distant stars, despite their great number, were too faint to be seen? At its birth, the riddle of darkness at night received what seemed a perfectly sensible answer,* one which was accepted by many astronomers who followed.

Thomas Digges's innovative contribution to the riddle of darkness is generally overlooked by astronomers and historians of science, perhaps because he found nothing paradoxical in the subject. Without doubt he originated the riddle, for he was the first person to appreciate that the dark gaps between visible stars call for an explanation.[9]

All Worlds Alike

The lapsed Dominican monk Giordano Bruno was much influenced by the idea of infinity then in the air. While living in England from 1583 to 1585 Bruno wrote *The Infinite Universe*. In the following years he wrote several works, of which *De Immenso* (1591) was the last and most important. Undoubtedly he had read Digges's "Perfit Description," and more than once he referred to the Epicurean system portrayed in *De Rerum Natura* and to the centerless and edgeless universe discussed in Nicholas of Cusa's *Learned Ignorance*.[10]

Whereas Digges still retained central symmetry in a universe without an outer edge, Bruno boldly abolished all trace of cosmic symmetry, both anthropocentric and heliocentric. He promoted a centerless uni-

*Entry 2 in the table "Proposed Solutions."

verse obeying the cosmological principle that all places are alike. Wherever he journeyed, he proclaimed with crusading zeal the message of an immense universe of countless stars encircled by inhabited worlds. "Innumerable celestial bodies, stars, globes, suns and earths may be sensibly perceived therein by us, and an infinite number of them may be inferred by our own reason. The universe, immense and infinite, is the complex of this space and all the bodies contained therein." Divine plenitude, not astronomy, had at last triumphed, and a boundless God had burst the bonds of the medieval system.

Giordano Bruno was charged with heresy in Venice by the Inquisition, imprisoned in Rome, and after years of torment taken to the Square of Flowers and burned at the stake in the early hours on Thursday, 16th of February, 1600.

In that same year William Gilbert, president of the College of Surgeons in London and the most distinguished Elizabethan scientist, published his book *The Magnet,* which differentiated between electricity and magnetism (he invented the word *electricity*) and showed that the Earth is a huge magnet, with north and south magnetic poles. Like many natural philosophers at this time, he accepted the Sun as the center of the Solar System, and promoted the idea of a plurality of inhabited worlds (Figure 3.7). Obviously he was influenced by Thomas Digges and probably by Bruno. The planets are at unequal distances from the Earth, he said, and in the same way, so are the stars. "How far removed from the Earth must be the most widely separated stars, and at a distance transcending all sight, all skill, all thought!"

> It is evident then that all the heavenly bodies, set as if in destined places, are there formed unto spheres, that they tend to their own centers, and that round them there is a confluence of all their parts. And if they have motion, that motion will rather be that of each round its own center, as that of the Earth is; or a forward movement of the center in the orbit, as that of the Moon.[11]

We cannot help but contrast the fate of Gilbert with that of Bruno; he was knighted by the queen and made her physician. In latitudinarian England, unlike catholic countries, the new ideas in astronomy were given free rein.

All Coherence Gone

The conceptual innovations of the late Middle Ages and the sixteenth century, which later culminated in the cosmic systems of the seventeenth and eighteenth centuries, were fourfold:

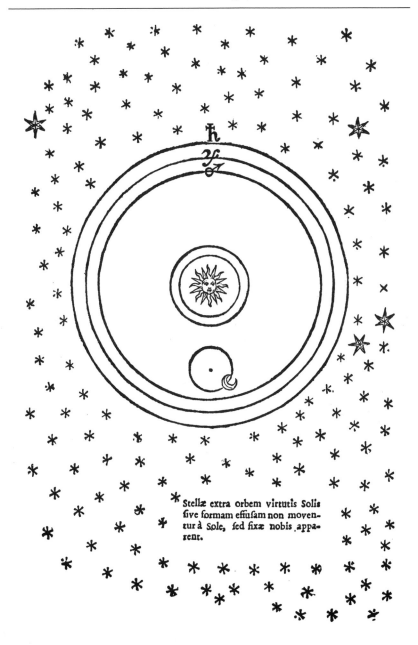

Stellæ extra orbem virtutis Solis five formam effusam non moventur à Sole, fed fixæ nobis apparent.

3.7   William Gilbert's cosmographical diagram in his posthumously published *New Philosophy.* (Reproduced as figure 9 in Dorothea Singer, *Giordano Bruno: His Life and Thought.*)

(1) The acceptance of the idea that living creatures inhabit the other planets of the celestial spheres.

(2) The erosion of the Aristotelian-medieval outer starry sphere and the dispersal of the stars either in a starry cosmos, beyond which lies an infinite void (the Stoic system), or throughout an infinite space (the Epicurean system).

(3) The realization that the fixed stars are possibly suns like our Sun.

(4) The realization that the fixed stars are possibly solar systems like our Solar System, with inhabited planets.

The new and startling ideas in astronomy inspired playwrights and poets of the sixteenth and seventeenth centuries in reformation countries, particularly in England, and much less noticeably in counterreformation countries.

The popular "Perfit Description" ran through many editions, greatly influencing Christopher Marlowe, John Donne, and John Milton but having little effect on William Shakespeare, "bounded in a nutshell," curiously unconcerned by the immensity of the new heavens and by the transformation from a geocentric to an acentric system. Shakespeare lived in a world of time.[12] The closest he came to our theme of cosmic darkness was in *As You Like It*, when the innocent shepherd Corin, asked by the clown Touchstone the extent of his philosophy, responded, "A great cause of the night is lack of the Sun."

Marlowe, on the other hand, lived in a world of space. He perceived immense unexplored astronomical vistas and exulted in the new vision. Donne surveyed "the new philosophy calling all in doubt," lamenting in the *Anatomy of the World*,

> The Sun is lost, and the earth, and no man's wit
> Can well direct him where to look for it.
> And freely men confess that this world's spent,
> When in the Planets, and the Firmament
> They seek so many new; then see that this
> Is crumbled out again to his Atomies.
> 'Tis all in pieces, all coherence gone;
> All just supply, and all Relation.

Milton, roaming the world of *Paradise Lost*, saw the "dark, unbottomed, infinite Abyss," the "wide womb of uncreated night," and with lonely steps sought

> The secrets of the hoary Deep—a dark
> Illimitable ocean, without bound,
> Without dimension; where length, breadth, and highth,
> And time, and place, are lost.

But we run ahead, for Milton struggled to express the amazement of an age confronted with the confounding, if not incomprehensible, world revealed by the telescope.

The Starry Message

Far above these heavens which here we see,
Be others far exceeding these in light,
Not bounded, not corrupt, as these same be,
But infinite in largeness and in height,
Unmoving, incorrupt and spotless bright.

Edmund Spenser, "Hymn of Heavenly Beauty"

"GOD said, Let there be light: and there was light." The ancient belief that light fills the heavens persisted into the sixteenth century and even later. To people in the Middle Ages, the "bright blue firmament" was not just scattered sunlight in the upper atmosphere, as we nowadays understand it, but actually the supernal light of heaven that grew in brightness as the soul ascended the angelic stairway of the celestial spheres and approached the empyrean.[1] Thomas Bradwardine, who said that God existed everywhere in an infinite and eternal void and that God had created within the void a finite cosmos of finite age, believed that space and light in the created cosmos were coextensive. The invention of the telescope at the dawn of the scientific age destroyed the old belief in celestial light and plunged the heavens into darkness.

From the overthrow of Aristotelian science and geocentric cosmology, and from the ensuing enlargement of space and dispersion of the stars, emerged the Cartesian and Newtonian systems of natural philosophy. Only in the context of these new and superior cosmological systems could the riddle of darkness emerge into the open and reveal its paradoxical form. But both these systems owe much to the work of Galileo and Kepler—two of the most remarkable scientists who ever lived.

Hosts of New Stars

Galileo Galilei (Figure 4.1), born in Pisa in 1564, brought order to the legacy of medieval science and made unforgettable astronomical discov-

4.1   Galileo Galilei (1564–1642).

eries with his magnifying tube. While we do not known with certainty who invented the telescope, credit is often given to Hans Lippershey, a lens grinder in Holland who serendipidously produced mangified images by combining lenses, and applied for a patent on his invention in 1608.[2] The word "telescope" was coined by John Demisiani of Cephalonia and introduced in 1611 at a banquet in Rome given in honor of Galileo's astronomical discoveries.

In 1609, at age forty-five, Galileo heard of the invention of the telescope and hastily set to work constructing his own version, which consisted of "two glass lenses, both plane on one side while on the other side one was spherically convex and the other concave." With this instrument he surveyed the heavens and a year later electrified the world of learning by describing his observations in a small book, *The Starry Message*, published in Venice.[3]

Galileo discovered mountains on the Moon, four moons of Jupiter (Io, Europa, Ganymede, and Callisto), and hosts of previously unseen stars. The Milky Way, he said, presents an astonishing spectacle, and "upon whatever part of it the telescope is directed, a vast crowd of stars is immediately presented to view. Many of them are large and quite bright, while the number of smaller ones is quite beyond calculation."

Astronomers have classified stars into magnitudes since the time of Hipparchus in the second century B.C. The first magnitude comprises about a dozen of the brightest stars, and the sixth magnitude comprises all stars barely visible to the unaided eye; the first magnitude is one hundred times brighter than sixth magnitude.[4] With telescopes Galileo extended the magnitude range to fainter stars: "In addition to stars of the sixth magnitude," he wrote, "a host of other stars are perceived through the telescope which escape the naked eye; these are so numerous as almost to surpass belief." The hitherto unseen stars made visible by his telescope greatly outnumbered the known stars seen with the naked eye. Publication of these results brought fame to Galileo.

Later, when he left the university at Padua and became a professor at Pisa, he observed the phases of Venus and showed how they prove that Venus orbits the Sun and not the Earth; he studied the spots on the Sun, which indicated, like the mountains on the Moon, that celestial bodies are not altogether unblemished, as Aristotle had asserted, and from the motion of the sunspots he discovered the Sun's 27-day period of rotation.

At age sixty-eight Galileo wrote his masterpiece *Dialogue Concerning the Two Chief World Systems* (1632), in which he contrasted unfavorably the geocentric Ptolemaic system with the heliocentric Copernican system and poured ridicule on many Aristotelian ideas still cherished by academics and clerics.[5] This work, written in the vernacular tongue and not Latin, was widely read and caused much concern. The Inquisition forced this outspoken scientist to abjure his "errors and heresies," and a sentence of life imprisonment was reduced to house arrest at his home in Arcetri near Florence, lasting until his death in 1642 at age seventy-eight. During his last years, despite infirmities, he wrote the important

Two New Sciences, which was surreptitiously published in Holland in 1638. His *Dialogue Concerning the Two Chief World Systems* remained on the *Index Librorum Prohibitorum* (*Index of Prohibited Books*) until the first half of the nineteenth century.

The Wizard Astronomer

Johannes Kepler (Figure 4.2), born in Württemberg, Germany, in 1571, the son of a soldier of fortune who deserted his family, suffered all his life from physical weaknesses and poor eyes. His brilliance in mathematics became evident while a youth at the University of Tübingen, and as a young man he taught mathematics and astronomy at Graz in Austria. Enthusiastically he embraced the heliocentric system of Copernicus and accepted Gilbert's idea that the Sun controls the planets by magnetic forces (Figure 4.3). He founded the science of optics, distinguished between light and vision, explained how the eye functions, and in later years developed the optical theory of the microscope and telescope. Kepler had mystical characteristics; like Pythagoras, he believed in the divinity of mathematical relations, and he worshipped a radiant being, symbolized by the Sun, which orchestrated the harmony of the spheres.

When Tycho Brahe, a Dane and the foremost astronomer of his time, died in 1601 in Prague, Kepler inherited from Tycho the title of Imperial Mathematician of the Holy Roman Empire. Imperial mathematician meant little more than court astrologer, and Kepler's official function was to cast horoscopes for state dignitaries and compile almanacs for guidance of the public. This was not an onerous duty, nor well-paid, and he devoted his time to more scientific pursuits when not distracted by poverty, illness, and family tragedy (for example, at one period he defended his mother in the courts against a charge of being a witch).

Kepler inherited, in addition, Tycho's recorded careful astronomical observations of the planets. From these observations, made with the utmost precision possible before the invention of the telescope, Kepler derived his famous three laws of planetary motion. The first law states that a planet moves in an elliptical orbit, with the Sun at the focus of the ellipse. The second law states that a straight line from the Sun to the moving planet sweeps out equal areas in equal intervals of time. He announced these first two laws in *The New Astronomy* (1609), and their discovery, made before the introduction of the telescope, terminated the long history of planetary epicycles. Kepler's third law, introduced in *The Harmony of the Spheres* (1619), in his own words states, "The periodic

4.2   Johannes Kepler (1571–1630).

times of any two planets are to each other exactly as the cubes of the square roots of their median distances"; in other words, the square of the orbital period of a planet is proportional to the cube of the distance to the Sun (the length of the semimajor axis).

Kepler passionately believed in a finite universe. In *The Mysterious*

Universe (1596) he argued that the Sun, the source of a magnetic force driving the planets, acts as the hub of the universe. Beyond the wheeling planets stand the fixed stars at the rim of the cosmos. We must reject, he declared, the thought of a limitless universe filled with stars all similar to the Sun; this awful thought dethrones and banishes the Sun into a wilderness of indeterminate space (Figure 4.4).

In *The New Star* (1605) Kepler again vigorously resisted the idea of an infinite sea of stars: "This very cogitation carries with it I don't know

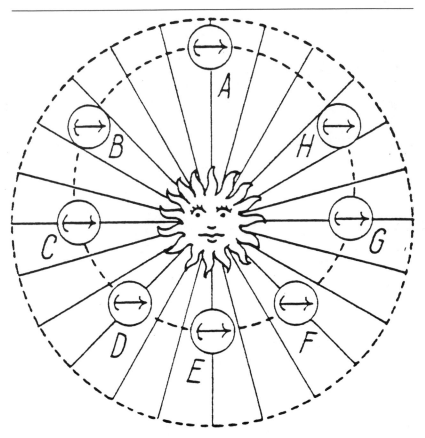

4.3   The Sun influencing the motion of the planets. This happy-looking symbol of the Sun is used several times by Kepler in his *Epitome of Copernican Astronomy* (1618). In this work he wrote, "I have built all Astronomy on the Copernican Hypothesis of the World; the Observations of Tycho Brahe; and the Magnetic Philosophy of the Englishman William Gilbert."

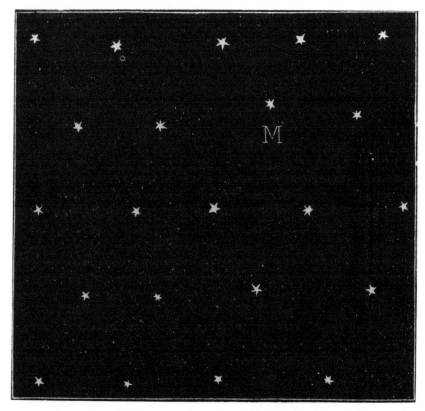

4.4 Kepler's M figure in *Epitome of Copernican Astronomy*. In an endless universe where everything looks much the same, said Kepler, our Sun would be
no more than an undistinguished star such as that labeled M.

what secret, hidden horror; indeed one finds oneself wandering in this
immensity, to which are denied limits and center and therefore all determinate places." Even to the most casual observer, he said, it must
seem obvious that the stars are not uniformly distributed. Between the
Sun and the fixed stars lies a great hollow, and the distant Milky Way
encloses us in an uninterrupted circle, holding us in the middle, and
"both the Milky Way and the fixed stars play the role of extremities.
They limit this our space, and in turn are limited on the exterior. Is it,
indeed, credible that, having a limit on this side, they extend on the
other side to infinity?"[6]
Kepler raised two potent objections against the idea of an endless

starry universe. The second objection relates to the main theme of our story, and we shall discuss it in the next section. The first objection, encountered in *The New Astronomy,* states that stars seem much the same size, although they differ in the magnitude of their brightness. Hence, concluded Kepler, stars cannot lie at different distances, but all lie at much the same distance, and have different intrinsic brightnesses.

The important point raised by Kepler was not fully explained until the nineteenth century. Rays of light are slightly scattered, or *diffracted,* when entering the pupil of the eye or the aperture of a telescope. Consider first the eye. Diffraction of light entering the pupil prevents the eye from resolving angles much less than about one arc minute. (One degree of arc contains sixty minutes, and each minute of arc contains sixty seconds.) A golf ball 150 meters away subtends an angle of one arc minute. (Actually, the distance is 147 meters for the American golf ball and 141 meters for the British golf ball. A table-tennis ball subtends the same angle at 131 meters.) A golf ball at much greater distance appears to have the same size as at 150 meters.

Diffraction—the deflection of light on passing through an aperture—increases with wavelength, thus accounting for the colorful prismatic effects in haloes seen about brightly lit objects. We must note also that the amount of diffraction diminishes as the aperture increases in size. In telescopes, because of their larger apertures, the diffraction is much less. As a rough guide, we can say that the limit of resolution on a photographic plate in a good telescope is in the vicinity of one arc second. A golf ball at nine kilometers (5½ miles) subtends an angle of one arc second. When seen by eye through a telescope, the image of the golf ball must be magnified more than sixty times to resolve its detail. The nearest stars subtend angles a thousand times smaller than one second of arc, and all stars, however distant, appear the same size when seen by the unaided eye and with the telescope. (This argument ignores other considerations such as the fine structure of the retina and the blurring caused by the atmosphere.)[7]

The Dark Night Sky

In 1610 a copy of Galileo's *Starry Message* reached Kepler, and after a few days he dashed off a letter in response to Galileo's request for the opinion of the imperial mathematician. A month later Kepler published this letter as a pamphlet entitled *Conversation with the Starry Messenger.*[8] In his reply Kepler referred to the moons of Jupiter as "satellites," from a Latin term denoting the obsequious attendants of an influential person.

Galileo's discovery of hosts of new stars at first alarmed Kepler. How could this be possible when the heliocentric universe consists of only a finite number of stars? Did this discovery mean that most stars are too far away to be seen without a telescope? No, because most likely the faint stars seen in the telescope are smaller bodies at roughly the same distance as the stars visible to the naked eye. The large number of new stars discovered by Galileo surely confirms that most are in fact much smaller than the Sun.

Kepler wrote in *Conversation with the Starry Messenger,* "You do not hesitate to declare that there are visible over 10,000 stars. The more there are, and the more crowded they are, the stronger becomes my argument against the infinity of the universe, as set forth in my book *The New Star.*" He then made the following important remark:

> Suppose that we took only 1,000 fixed stars, none of them larger than 1 minute of arc (yet the majority in the catalogues are larger). If these were all merged in a single round surface, they would equal (and even surpass) the diameter of the sun. If the little disks of 10,000 stars are fused into one, how much more will their visible size exceed the apparent disk of the sun? If this is true, and if they are suns having the same nature as our sun, why do not these suns collectively outdistance our sun in brilliance? Why do they all together transmit so dim a light to the most accessible places?

Here we see Kepler, with groping thoughts, formulating his second and more cogent objection to the idea of an endless starry universe. He argued that the large number of new stars discovered by Galileo proves that the fixed stars must be smaller and less luminous than the Sun; otherwise the celestial vault would be more luminous than the Sun. He failed to realize that the size of stars is unimportant. (Trees in a forest form a continuous background no matter how thin their trunks.) Nonetheless, he had stumbled on a remarkable argument: The larger the star-filled universe, the more the stars must cover the sky. At last the riddle of a dark starlit sky has come into focus.

Kepler thought the sky was dark at night because the universe simply contained too few stars to cover the whole sky. The finiteness of a bounded universe explained why the stars were too few (Figure 4.5). In his last major work, *Epitome of Copernican Astronomy* (1618), he said the world of stars was "enclosed and circumscribed as by a wall or a vault,"[9] and in this sense was similar to the Aristotelian system.*

---

*Entry 3 in the table "Proposed Solutions."

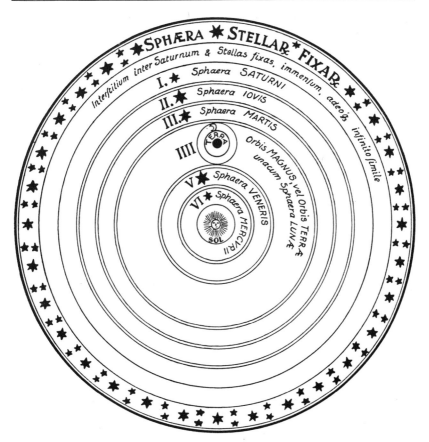

4.5 Kepler's diagram of a bounded Copernican universe in the *Prodomus to the Mysterious Universe* (1596). In *Epitome of Copernican Astronomy*, his last major work, he maintained his belief in a bounded universe.

Kepler rejected the Epicurean system. He also rejected the Stoic system. Endless empty space beyond the starry cosmos, "as such, neither *is*—how, indeed could it *be* if it is nothing?—nor has it been created by God, who assuredly has created the world out of nothing, but did not start by creating 'nothing.'"[10] Kepler shared the prevailing belief that human beings are privileged above all other creatures in the cosmic scheme of things. At the end of *Conversation with the Starry Messenger* he asked, "If there are globes in the heaven similar to our earth . . . how can all things be for man's sake? How can we be the masters of God's handiwork?" Creatures on other worlds who served no purpose in human life had no conceivable place in the divine plan.

Despite Kepler's potent objections to a boundless universe, by the middle of the seventeenth century the Aristotelian system sprawled in ruins, its outer walls torn down. Above the skyline could be seen the towering Cartesian system, soon to be overshadowed by the skyscraping Newtonian system.

PART TWO

★ THE ★
RIDDLE
DEVELOPS

The Cartesian System

> But we must not try to dispute about the infinite, but just consider that all that in which we find no limits is indefinite, such as the extension of the world, the divisibility of its parts, the number of stars, etc.

Descartes, *Principles of Philosophy*

RENÉ DESCARTES was a youth in the Jesuit college at La Fleche in 1610 when Kepler enunciated the riddle of darkness to puzzle all who disagreed with his belief in a finite universe.

Descartes (Figure 5.1), a famed mathematician and philosopher of extraordinary originality, was born at La Haye (near Tours) in France in 1596. Like Newton, he never married; unlike Newton, who lived to eighty-four, he died prematurely at fifty-four. He was summoned to Stockholm by Queen Christina for the purpose of elevating the intellectual level of the Swedish court, an assignment requiring in addition that he instruct her in philosophy three times a week at 5 a.m. in the middle of winter. This proved too much for Descartes—he normally preferred to work in bed and was never very robust—and in 1650 he died of pneumonia.

Even while Galileo was suffering the condemnation of the Holy Office in his declining years, Descartes, taking refuge in protestant Holland, was formulating his new philosophy in *Discourse on the Method* (1637). Guided by the method of "rightly conducting the reason" and the principle that what is reasonable must be true, he constructed a grand and sweeping philosophy. Its most enduring features have turned out to be the mathematization of the physical sciences and the dichotomy of body and mind. In his *Principles of Philosophy* (1644), addressed to his pupil and the woman he loved, the lively Elizabeth of Bohemia (Princess Palatine and granddaughter of Mary, Queen of Scots), Descartes introduced a system of natural philosophy more innovative than any since classical antiquity.[1]

5.1   René Descartes (1596–1650).

Only God could be infinite in a real sense, said Descartes, and he would refer to the spatial extent of material things as "indefinite rather than infinite in order to reserve to God alone the name of infinite." Space in Descartes's system extended in all directions to indefinite distances, and continuous matter pervaded the whole of space. The fully mechanized external world of matter and motion obeyed natural laws free of supernatural intervention. In the beginning God established the natural laws, said Descartes, and "we may well believe, without doing outrage to the miracle of creation, that by this means alone all things which are purely material might in course of time have become such as we observe them to be at present."

Descartes promoted the principle of action by direct contact. Bodies move in straight lines unless compelled by forces to follow curved lines, he said, and the applied forces, pushing and pulling, are pressures and tensions due to adjacent bodies and elements of fluid. Whirlpools and vortices on all scales, pressing against one another, account for the rotating and revolving motions of stars and their planets. Descartes apparently remained unaware of Kepler's three laws of planetary motion and did not, fortunately for him, have to explain these laws by the action of vortical fluids in his whirligig system of swirling fluids and gyrating worlds (Figure 5.2).

Reason persuades us, said Descartes, that space by itself, being nothing, has no extension. How can mere space alone, which is empty and nothing, have length, breadth, and height? Only matter has the property of extension, and space cannot exist where there is no matter. Matter in many forms—dense in planetary interiors and rarified in interplanetary and interstellar space—exists everywhere, and a vacuum, repugnant to reason, exists nowhere. The nonexistence of space by itself remained the cardinal idea in the Cartesian system of natural philosophy.

From this idea sprang Descartes's firm belief in the impossibility of the atomic theory. The principle of action by direct contact required that matter be continuous, free from voids, and nonatomic. Atoms, said Descartes, are separated by voids, and voids, being repugnant to reason, are physically impossible. Consider: "atoms" when separated by nothing have direct contact with one another, and hence are not atoms but only the parts of a continuous whole; furthermore, "however small the parts are supposed to be, yet because they are necessarily extended, we are always able in thought to divide any one of them into two or more parts; and thus we know that they are divisible."

Evangelista Torricelli, Galileo's companion during the last months of his life, investigated nature's abhorrence of a vacuum. In 1643 he succeeded, with surprising ease, in creating a vacuum in a glass tube above a column of mercury of height thirty inches. He poured mercury into a glass tube sealed at one end, then inverted the tube, and dipped the

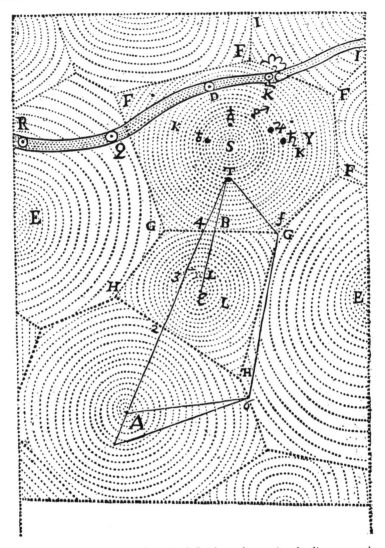

5.2 The Cartesian system of vortical fluids and gyrating bodies, reproduced from Descartes's *The World, or a Treatise on Light* (1636). Each vortex denotes a solar system in a continuous and endless expanse of solar systems. The centers of vortices, such as those indicated by S, E, and A, are actually stars made luminous by the central churning motion in the vortices. The dotted lines indicate the paths of fluid elements that compose the vortices. The body C, shown crossing the top of the diagram, is a comet moving too fast to be retained by any vortical solar system.

open end in a bowl of mercury. Torricelli performed various tests and correctly concluded that a vacuum exists above the mercury column, and explained how the height of the column measures the weight (or rather pressure) of the atmosphere. He noticed that the height varies from day to day because of atmospheric changes, and he was therefore the inventor of the mercury barometer (Figure 5.3).

This evidence against the Cartesian system caused a certain amount of concern. Many Cartesians throughout the seventeenth century remained unconvinced, arguing that Torricelli's space above the mercury column still contained air, but greatly rarified, and was not a perfect vacuum.

The Rise and Fall of Cartesian Cosmology

The Cartesians believed in a universe of indefinite extent, in which all things are pushed and pulled by forces acting in direct contact. All the nonsense of forces acting at a distance, of atoms and the vacuum, and (except for Galileo and Christiaan Huygens) of light traveling at finite speed had no place in their elegant and perfectly rational scheme of things.

Despite strenuous opposition from clerics and academics steeped in otherworldly doctrine, the exhilarating new Cartesian philosophy spread rapidly, capturing the imagination of freethinkers. It signposted the way to the Age of Reason of the eighteenth century; also it influenced the course of science and theology in England and stimulated the rise of the Newtonian system.

At first English liberal theologians and philosophers viewed Descartes as a savior. But soon his philosophy received more criticism than praise. The Cambridge theologian Henry More, initially impressed with the vision of a universe of natural laws, in later years, aghast at the implications of Cartesian materialism, returned to the old idea that space exists without matter by virtue of the presence of spirit.[2] More believed in a Stoic finite material cosmos immersed in an infinite extramundane space. Quite possibly More's ideas on the nature of space influenced Isaac Newton in his early years at Cambridge.

Robert Boyle, Christopher Wren, Robert Hooke, Isaac Newton, Edmund Halley, and other natural philosophers of England rejected the Cartesian blueprint of dead matter ruled by purely mechanistic laws. The Newtonians retained the astrological emanations and astral forces of the medieval universe, whereas Descartes, Huygens, Gottfried Leib-

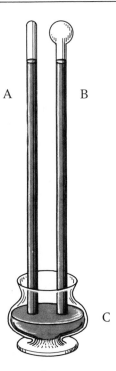

5.3 Evangelista Torricelli's vacuum above a mercury column. When the two glass tubes A and B were "filled with quicksilver," wrote Torricelli in 1644, and "their mouths stopped with the finger, and then turned upside-down in a vase C that had some quicksilver in it, they were seen to empty themselves." The height of the mercury column was independent of the shape of the glass tube, thus showing, argued Torricelli, that the force holding up the mercury did not reside in the vacuum. (See W. E. Knowles Middleton, *The History of the Barometer*.)

niz, Bernard de Fontenelle, and other natural philosophers of continental Europe would have nothing to do with this legacy of astral influences acting across wide empty spaces. Whatever Descartes said, Newton later said the opposite. The God of infinite space and eternal time not only had created the material universe, according to Newton, but also was manifest in its wonder and glory, and was constantly engaged in maintaining it in working order.

After Kepler, in the upheaval of cosmic reconstruction, the riddle of darkness at first received relatively little direct attention. We find various hints, but not until the work of Edmund Halley does the riddle reach a climax that reveals its paradoxical nature. One very obvious solution, however, was proposed before the birth of the Newtonian system by Otto von Guericke, an outspoken critic of Cartesian philosophy.

The Magdeburg Experiments

Otto von Guericke (Figure 5.4), a German soldier, engineer, and scientist, toward the end of the Thirty Years' War returned to Magdeburg, the town of his birth, and helped to rebuild it. He became mayor from 1646

5.4   Otto von Guericke (1602–1686).

5.5 Otto von Guericke in 1650 constructed the first air pump and became famous for performing spectacular experiments. The most famous of his demonstrations was performed for the first time at Magdeburg in 1657. Two copper hemispheres, held together by a vacuum, could not be pulled apart by teams of horses. In the experiment shown in this illustration Guericke measured the atmospheric pressure holding together the two halves of an evacuated sphere.

to 1676 (also thirty years), and in this period performed spectacular experiments with large evacuated vessels (Figure 5.5).[3] He demonstrated that candles cannot burn and animals cannot live in a vacuum. He also showed that a vacuum transmits light but not sound. His celebrated experiments disproved the old notion that a free-falling body in a vacuum would move at infinite speed. This idea was based on the prevailing Aristotelian belief that bodies move only when compelled, and the less resistive the medium the faster they must move.

The astonishing mayor experimented with static electricity and was the first to suggest that comets, as members of the Solar System, return periodically to the Sun's neighborhood.[4] In 1672, at the age of seventy, he published his ideas and experimental results in *The New Magdeburg Experiments on Void Space*.

Only God and space can be infinite, Guericke said, and though the starry cosmos may be immense, it is nonetheless finite in size. He used the forest analogy and said that stars are not limitless in number like trees in an endless forest. Guericke believed in a finite Stoic cosmos, and thought that the gaps between stars reveal to us the emptiness and darkness of an extracosmic void (Figure 5.6). He is the first on record to suggest that the Stoic system solved the riddle of darkness.*

The doctrine that a wise Contriver had created within an infinite and eternal space a material world of finite size and age was widely accepted in Guericke's day. Natural philosophers generally favored the Stoic system, and conceivably many assumed, like Guericke, that the dark gaps between stars gave evidence of the extracosmic void. The riddle raised by Digges and Kepler had an obvious solution for those, unlike Descartes, who believed in the Stoic cosmos—a solution so obvious that it seemed scarcely worth mentioning.

Light Crisscrossing the Heavens

In the springtime of science—the seventeenth century—when René Descartes declared that rays of light travel in straight lines and cross one another without hindrance, and the Lucasian Professor of Mathematics at Cambridge, Dr. Isaac Newton, expounded a theory that sunlight consists of rays of different colors unequally refracted,[5] who better can we turn to than Robert Hooke, the ingenious Curator of Experiments at the Royal Society who blended Cartesian and neo-Newtonian natural philosophies, to see the properties of light brought into sharper focus? "Probably the most inventive man who ever lived," writes Edward Andrade of Robert Hooke.[6] Born at Freshwater on the Isle of Wight in 1635, the son of a clergyman, he suffered from physical handicaps and continual illness. He mixed at Oxford with the natural philosophers who later formed the nucleus of the Royal Society in London. On Jan-

---

*Entry 4 in the table "Proposed Solutions."

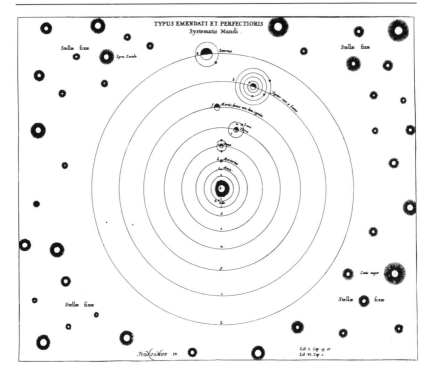

5.6 Otto von Guericke's system of the world in *New Magdeburg Experiments on Void Space* (1672), showing the dispersal of the fixed stars. He wrote, "Although many stars cannot be seen we should not form the opinion that they do not exist. A forest does not end where individual trees cannot be seen any farther, and probably no more stars will be seen shining after passing through the whole stellar realm made visible with the telescope" (book 7, chap. 2, p. 57). Guericke strenuously opposed Descartes's arguments against the vacuum. He believed in a finite starry cosmos surrounded by an infinite void, as in the Stoic system, and said the sky is dark at night because we look between the stars and see the starless void beyond.

uary 20, 1665, Samuel Pepys went to his bookseller and "took home Hook's book on microscopy, a most excellent piece, and of which I am very proud." The response in scientific circles to Hooke's illustrated *Micrographia*, which explored the microscopic world, rivaled that of Galileo's *Starry Message*.

In his "Lectures of light explicating its nature, properties, and effects," delivered about the beginning of 1680, Hooke considered rays of light crisscrossing the heavens. Each luminous point radiates rays spherically, he said, and these rays travel through the "Diaphanous Medium" to unlimited distances: "The Radiations of Light I have formerly Explained are continually Propagated from every Luminous Object, and from every Point of every Luminous Object . . . So that as there are infinite of these radiating Points in the World, so from every one of those infinite Points there are infinite of those Radiations." Rays from an infinity of luminous points enter through the pupil into the "Black Hole of the Eye," making the eye a microcosm of the universe: "it having a distinct Point within it self for every distinct Point without it self in the Universe; and when a Hemisphere of the Heavens is open to its view, it has a Hemisphere within it self, wherein there are as many Respective Points for Reception of Radiations as there are differing Points for emission of Radiations."[7]

Hooke thought correctly, as Digges had done a hundred years earlier, that the light rays received from a star at very great distance are too weak to register an impression on the eye. But he failed to consider the accumulative effect of numerous weak rays from many stars. The eye cannot detect a weak ray emitted by a single atom, but can detect the combined rays from many atoms (Figure 5.7).[8]

Nowhere did Hooke say that he actually believed in an infinite star-filled universe. On the extent of space itself he said, "Whether it be finite or infinite, the Vastness of it is so great that it exceeds our Imagination to conceive of it truly as it is." The extent of the stars in space must also be vast, "since we find that still longer and better Telescopes do discover to us smaller and smaller fixt Stars, which in Probability are farther and farther removed from us." Possibly Hooke thought, like Guericke, that the starry cosmos was limited and less than the extent of space, thus explaining the darkness of the night sky.

Bernard de Fontenelle, born in Rouen in 1657, author and scientist, a man of wit and humor, a full-blooded Cartesian gracing the salons of great ladies in the gay social life of Paris, died only a month short of reaching the age of one-hundred. By skillfully avoiding theistic doctrine he acted as a forerunner of the deists in the Age of Reason. (Theism is the belief that God created and runs the universe, deism the belief that God created a self-running universe.) Fontenelle, who had a talent for lucid exposition, popularized science by making it both interesting and intelligible to a wide audience. At age twenty-nine, in *Conversations on the Plurality of Worlds*, he gave an explanation of the Copernican system that succeeded in converting many persons previously unconvinced. He

5.7   The excited atoms in a candle flame emit pulses of light much too feeble to be detected individually by the eye. Their collective effect, however, accounts for the visible incandescence of the flame. Digges, perhaps Hooke, and apparently Halley failed to appreciate the collective effect of many sources individually too feeble to be detected.

held imaginary conversations with a countess and vividly described the astronomical realm in terms of Cartesian vortices.

"You see that whiteness in the Sky," expatiated Fontenelle to the countess, "which some call the *milky way;* Can you imagine what that is? 'Tis nothing but an infinity of small Stars, not to be seen by our Eyes, because they are so very little; and they are sown so thick, one by another, that they seem to be one continu'd whiteness. I wish you had a Glass, to see this Ant-hill of Stars, and this Cluster of Worlds, if I may so call them."[9] If you lived on one of these island worlds in the Milky Way, he said, you would see

your Heaven shine bright with an infinite number of Fires, close to one another, and but a little distant from you; so that tho' you should lose the light of your own particular Sun, yet there would still remain visible Suns enough beside your own to make your Night as light as Day, at least, the difference would hardly be perceiv'd, for the truth is, you would never have any Night at all. The Inhabitants of these Worlds accustom'd to perpetual Brightness would be strangely astonish'd, if they should be told that there are a miserable sort of People, who where they live, have very dark Nights, and when 'tis Day with them, they never see more than one Sun; certainly they would think Nature had very little kindness for us, and would tremble with horrour, to think what a sad condition we are in.

Fontenelle performed the creditable service of describing what happens when visible stars, filling sections of the sky, turn night into day and banish darkness from the heavens.

Christiaan Huygens of Holland, an eminent natural philosopher of the seventeenth century, discovered the rings of Saturn and developed a wave theory of light. In *Celestial Worlds Discover'd*, posthumously published in 1698, he wrote, "But what God has been pleas'd to place beyond the Region of the Stars is as much above our Knowledge as it is our Habitation. Or what if beyond such a determinate space he has left an infinite Vacuum to show how inconsiderable is all that he has made, to what his Power could, had he so pleas'd, have produc'd?[10] In this passage Huygens expressed a variant form of Cartesian philosophy having much in common with the Stoic system. He imagined God coextensive with infinite mysterious space and regarded the starry cosmos as finite and determinate in size.

Guericke, perhaps also Hooke, had supposed much the same as Huygens. Plausibly this is the reason why the riddle of cosmic darkness received little serious attention between Kepler's *Conversation with the Starry Messenger* and Halley's important contribution more than a century later.

Newton's Needles and Halley's Shells

The heavens are all his own, from the wide rule
Of whirling vortices and circling spheres
To their first great simplicity restored.
The schools astonished stood; but found it vain
To combat still with demonstration strong,
And, unawakened, dream beneath the blaze
Of truth. At once their pleasing visions fled,
With the gay shadows of the morning mixed,
When Newton rose, our philosophic sun!

James Thomson, "Newton"

★

THE riddle of cosmic darkness became far more puzzling with the rise of the Cartesian and Newtonian systems, and by 1721 it had become, in Edmund Halley's words, a "metaphysical paradox," challenging all who believed in a universe of endless stars.

The Cartesian system combined Aristotelian and Epicurean ingredients: Aristotelian because it rejected the vacuum and atomicity of matter, and Epicurean because it accepted the unlimited extent of space. The Newtonian system in its preliminary form combined Stoic and Epicurean ingredients: Stoic because it rejected an unlimited expanse of stars, and Epicurean because it accepted the vacuum and the atomicity of matter (Figure 6.1). Later, soon after Newton had developed the theory of universal gravity, the Newtonian system became almost entirely Epicurean, and the Stoic finite cosmos was abandoned in favor of an Epicurean endless distribution of stars.

Henry More and Isaac Newton (Figure 6.2) believed that divine spirit pervaded the universe. Where there was no matter, spirit alone sufficed to endow space with extension. To say that space cannot exist where

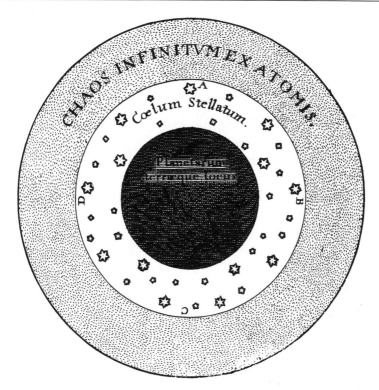

6.1 The "world-system of the ancients," according to Edward Sherburne's translation (London, 1675) of *The Sphere* by Manilius. This illustration combines elements from the Epicurean and Stoic systems and plausibly represents Newton's view of the universe when he wrote *De gravitatione* sometime between 1666 and 1668, during his early years at Cambridge.

there is no matter denied the presence of spirit, hence denied the presence of God in the universe. The Newtonians found nothing repugnant about a vacuum and the atomicity of matter, and Boyle and Hooke developed air pumps, experimented with evacuated vessels, and used atomic ideas to explain the properties of matter. The thought of astral forces acting across interplanetary voids and atomic forces acting across interatomic voids failed to deter the Newtonians, armed with the concept of space pervaded by spirit.[1]

Newton formulated his ideas on space in his early years at Cambridge in response to Descartes's *Principles of Philosophy.* Sometime between 1666 and 1668, in an unpublished manuscript referred to by its opening

6.2　Isaac Newton (1642–1727).

words *De gravitatione,* Newton wrote, an "infinite and eternal" divine power "extends infinitely in all directions" and furthermore "is eternal in duration."[2] Descartes claimed that where there is no matter there can be no space; on the contrary, said Newton, space by virtue of omnipresent spirit exists in the absence of matter. Descartes claimed that matter extends indefinitely; on the contrary, said Newton, God has created in infinite mysterious space a material system of finite extent.

We can imagine, wrote the young Newton, "that there is nothing in space, yet we cannot think that space does not exist, just as we cannot think that there is no duration, even though it would be possible to suppose that nothing whatever endures. This is manifest from the spaces beyond the world, which we must suppose to exist (since we imagine the world to be finite)." At this period in his life Newton believed in a Stoic cosmos. In the second half of the seventeenth century we see on one side the Cartesian system of materialized space equipped with natural laws governing a world of dead matter, and on the other the Newtonian system of spiritualized space also equipped with natural laws, but under providential guidance, governing a world of matter enlivened by atomic forces.

Newton's ideas on the nature of space changed surprisingly little in his lifetime. His ideas on the nature of the universe, however, changed considerably after he had developed the theory of universal gravity; he abandoned the Stoic finite cosmos, substituting in its place an infinite star-filled Epicurean system.

The Principia

In *Mathematical Principles of Natural Philosophy* (the *Principia*) published in 1687, Isaac Newton at the age of forty-five formulated rigorous laws of motion and demonstrated that all bodies in the universe influence one another with gravitational forces proportional to their masses divided by the square of their separating distances, and that these forces control the movements of celestial bodies. He succeeded in explaining Kepler's three laws of planetary motion, the orbits of planets, satellites, and comets, the twice-daily terrestrial tides, the precession of the equinoxes, the rotational bulge of the Earth, and whatever else at the time seemed significant in a dynamic universe. Edmund Halley, who encouraged Newton to write the *Principia* and paid for and edited its production, wrote in a review, "It may be justly said that so many and so Valuable Philosophical Truths, as are herein discovered and put past

Dispute, were never yet owing to the Capacity and Industry of any one Man."[3]

Newton made numerous revisions to the *Principia,* and the three editions in 1687, 1713, and 1726 reveal very little of his thoughts regarding the extent of the starry universe. For information on this subject we must look elsewhere, in his correspondence and unpublished papers.

The Bentley Correspondence

Probably Newton's correspondence with the young clergyman Richard Bentley caused him to abandon the Stoic cosmos. Robert Boyle in his will left an endowment to provide sufficient income for an annual lectureship to combat the atheism widely professed by wits in coffeehouses and taverns. The trustees of the endowment selected the scholarly clergyman Richard Bentley to give, in 1692, the first series of Boyle Lectures.[4] No one better than the immensely erudite Bentley could have been chosen for the task of confuting atheism by appeal to reason rather than to faith.

Armed with Newton's "sublime discoveries," Bentley argued that the laws of nature are insufficient to explain the workings of the universe and must occasionally be supplemented by supernatural acts of a divine power. After sending the lectures to the printer, he took the precaution of consulting Newton on some technical points so that last-minute changes could be made. Bentley's probing and disturbing inquiries jolted Newton into reformulating his cosmological ideas, and Newton's four letters to Bentley rank among the most important documents in the history of science.[5]

In the first letter Newton declared his approval of the lectures. To Bentley's query about the effect of gravity in a system of finite size composed of material bodies, he responded with the famous statements:

> As to your first Query, it seems to me, that if the matter of our Sun & Planets & ye matter of the Universe was eavenly scattered throughout all the heavens, & every particle had an innate gravity towards all the rest & the whole space throughout wch this matter was scattered was but finite: the matter on ye outside of this space would by its gravity tend toward all ye matter on the inside & by consequence fall down into ye middle of the whole space & there compose one great spherical mass. But if the matter was eavenly diffused through an infinite space, it would never convene into one mass but some of it convene into one mass & some into another so as to make an infinite number of great masses scattered at great dis-

tances from one another throughout all yt infinite space. And thus might ye Sun and Fixt stars be formed supposing the matter were of a lucid nature.

Newton expressed the opinion that a self-gravitating material system was necessarily infinite and unbounded, for, if finite and bounded, it lacked an equilibrium state and would collapse. Possibly Newton recalled that Lucretius had said much the same in *De Rerum Natura:* "Further, if all the space in the universe were shut in and confined on every side by definite boundaries, the supply of matter would already have accumulated by its own weight at the bottom, and nothing could happen under the dome of the sky—indeed, there would be no sky and no sunlight, since all the available matter would have settled down and would be lying in a heap throughout eternity."[6]

Newton and Bentley agreed that stars stretch away endlessly (as in the Epicurean system), for if the sidereal system were finite (as in the Stoic system) the stars would fall into the middle. In his second letter Newton said, "You argue that every particle of matter in an infinite space has an infinite quantity of matter on all sides & by consequence an infinite attraction every way & therefore must rest *in equilibrio* because all infinites are equal." He agreed with Bentley that providence had designed a universe of infinite extent in which uniformly distributed stars stand poised in unstable equilibrium like needles on their points (Figure 6.3). He then explained how finite forces may still remain when infinite forces are subtracted from one another. "When therefore I said that matter eavenly spread through all spaces would convene by its gravity into one or more great masses, I understand it of matter not resting in an accurate poise."

Apparently Bentley's remarks prompted Newton to draft and redraft a long treatment for inclusion in the second edition of the *Principia.*[7] In this treatment he attempted to show that astronomical observations supported the idea of a uniform distribution of stars. He assumed that stars resemble the Sun and arranged their distances according to their apparent magnitudes. Out of this inconclusive study came eventually the terse remark in the second edition, "The fixt Stars, every where promiscuously dispers'd in the heavens, by their contrary attractions destroy their mutual actions."

In the second letter Newton also wrote, "So then gravity may put ye planets into motion but without ye divine power it could never put then into such a Circulating motion as they have about ye Sun, & therefore

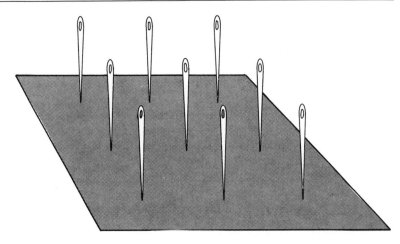

6.3 Newton and Bentley agreed that stars cannot form a finite system because they would all then fall into the middle. Instead, they must be uniformly distributed throughout infinite space. Gravitational forces would pull equally in all directions and each star would be in equilibrium. Both realized that this state of equilibrium was unstable. The stars, said Newton, are like an array of needles standing upright on their points, each ready to fall one way or another at the least disturbance.

for this as well as other reasons I am compelled to ascribe ye frame of this Systeme to an intelligent agent." The theme of an infinite universe and the role of providence runs through all four letters. Bentley wanted neither self-sufficient natural laws that explained everything nor ad hoc miracles at which atheists would scoff. He wanted Newtonian natural laws supplemented by indispensable providential acts, thus providing proof of God's existence. A Stoic system sustained solely by divine will provided no proof, but an Epicurean system in equilibrium according to natural laws, with planetary systems forming by gravitational collapse, offered ample scope for supplementary providential acts.

Newton wrote in the General Scholium of the second edition of the *Principia*,

> This most beautiful System of the Sun, Planets and Comets, could only proceed from the counsel and dominion of an intelligent and powerful being. And if the fixed Stars are the centres of other like systems, these being form'd by the like wise counsel, must be all subject to the dominion of One; especially, since the light of the fixed Stars is of the same nature with the light of the Sun, and from every system light passes into all the

other systems. And lest the systems of the fixed Stars should, by their gravity, fall on each other mutually, he hath placed those Systems at immense distances one from another.[8]

On the subject of atoms, Newton wrote in the *Opticks* (first published in 1704), "It seems probable to me that God in the Beginning form'd Matter in solid, massy, hard, impenetrable, moveable Particles, of such Sizes and Figures, and with such other Properties, and in such Proportion to Space, as most conduced to the End for which he form'd them."[9] The ancient atomist philosophy, stripped of every atheistic vestige, was thus made palatable to the Anglican Church.

Voltaire, taking refuge from incensed victims of his satirical pen, was resident in England in 1727 at the time of Newton's death. In one of his letters he wrote, "A Frenchman who arrives in London finds a great change in philosophy as in everything else. He left the world full, he finds it empty. In Paris one sees the universe composed of vortices of subtle matter. In London one finds nothing of this. In Paris it is the pressure of the Moon that causes the flux of the sea; in England it is the sea that gravitates toward the Moon . . . With your Cartesians, everything is done by an impulsion that nobody understands; with Mr. Newton it is an attraction whose cause is not better understood."[10] A few years later he again took refuge, this time at the chateau of Mme du Chatelet, who was interested in the sciences. While she translated Newton's *Principia* into French, Voltaire wrote *Elements of Sir Isaac Newton's Philosophy* (1736), which made Newtonian science fashionable in France.

"I Have Heard Urged"

Edmund Halley (Figure 6.4), born in Haggerston (later engulfed by the sprawl of London), felt attracted to astronomy while still a youth. At age twenty, after leaving Oxford before finishing his studies, he gained recognition for his astronomical observations made on the island of St. Helena in the southern hemisphere. He is celebrated for discovering the periodicity of Halley's comet. He was the first to discover a globular cluster of stars, and by noticing that the stars Arcturus, Procyon, and Sirius had changed their positions since ancient times, he was the first to discover that stars are not fixed in the heavens.[11] Halley's papers show signs of his growing interest in the subject of starlit skies. About luminous clouds in the heavens, he wrote in 1714, "In all these so vast

6.4   Edmund Halley (1656–1742).

Spaces it should seem there is a perpetual uninterrupted Day, which may furnish Matter of Speculation, as well as to the curious Naturalist as to the Astronomer."[12]

In 1721 at age sixty-four—the year after he became astronomer royal—Halley wrote two short papers (reproduced in Appendix 2) on the infinity of the universe.[13] In the first, "Of the infinity of the sphere of fix'd stars," he began with the comment, "The System of the World as it is now understood is taken to occupy the whole *Abyss* of *Space*, and to be as such actually infinite; and the appearance of the Sphere of Fixt Stars, still discovering smaller and smaller ones as you apply better Telescopes, seems to confirm this Doctrine." Furthermore, "Were the whole System finite . . . the whole would be surrounded on all sides with an infinite *inane*, and the superficial Stars would gravitate towards those near the center, and with an accelerated motion run into them, and in process of time coalesce and unite with them into one."

In these opening remarks he saw two objections to a finite starry cosmos surrounded by an infinite void (inane). First, every improvement in the telescope disclosed fainter and more distant stars, and hence in the dark gaps that separate visible stars hide invisible stars awaiting detection. Second, a finite material system lacked equilibrium and must collapse. For "if the whole be Infinite," he continued, "all the parts of it would be nearly *in equilibrio*, and consequently each fixt Star, being drawn by contrary Powers, would keep its place; or move till such time, as from such an *equilibrium*, it found its resting place; on which account some, perhaps, may think the Infinity of the Sphere of Fixt Stars no very precarious Postulate." Halley thought that the assumed equilibrium of an infinite and uniform universe was stable. Newton in his letters to Bentley was under no such illusion, and fully realized that the assumed equilibrium was unstable, like that of an array of needles standing on their points.

Halley continued: "Another Argument I have heard urged, that if the number of Fixt Stars were more than finite, the whole superficies of their apparent Sphere would be luminous, for that those shining Bodies would be more in number than there are Seconds of a Degree in the *area* of the whole Spherical Surface, which I think cannot be denied." The argument he had heard urged presumably stemmed from Kepler's work.[14] Though obscure in places, Halley's discussion in 1721 drew attention to the idea that an infinity of stars in an unbounded universe should cover the entire sky. The endless star-filled Newtonian universe thus brought into prominence the riddle of cosmic darkness and made compelling the covered-sky interpretation.

Why is the sky dark at night in an infinite, star-filled universe? Halley tried to give an answer. He said that the apparent luminosity of a star is proportional to the reciprocal of its distance squared: halving the distance of a star increases its apparent luminosity fourfold. By "an obvious *calculus*" he found that the separation between stars is proportional to the reciprocal of distance: halving the distance of stars doubles their separation. So far he is correct if he means the apparent separation between uniformly distributed stars. He deduced incorrectly, however, that the difference between these two effects—decrease in apparent luminosity as the square of distance and decrease in apparent separation as the distance—explained darkness at night.*

Perhaps realizing that this argument lacked clarity, Halley offered an alternate solution. The sky is dark at night, he said, because "the more remote stars, and those far short of the remotest, vanish even in the nicest Telescopes, by reason of their extream minuteness; so that, tho' it were true, that some such Stars are in such a place, yet their Beams, aided by any help yet known, are not sufficient to move our Sense; after the same manner as a small Telescopical fixt Star is by no means perceivable to the naked Eye." This argument, similar to that given by Digges 145 years previously, assumes correctly that rays from individual distant stars are much too weak to register an impression in the eye, but assumes incorrectly that the combination of rays from numerous stars is also too weak.

In the second paper, "Of the number, order, and light of the fix'd Stars," Halley referred to the "Metaphysical Paradox" he had raised in the first paper. He discussed stellar magnitudes, and the increase in the number of observed stars at each step in magnitude, but failed to correct the confusing errors in the first paper. From a remote star we receive, he concluded, "so small a pulse of Light, that it may well be questioned, whether the Eye, assisted with any artificial help, can be made sensible thereof."

An Infinity of Shells

Hitherto we have used the line-of-sight argument to show that the sky should blaze with starlight at every point in an endless universe filled with stars. The usual argument, introduced by Halley and much older than the line-of-sight argument, goes as follows (Figure 6.5): We con-

---

*Entry 5 in the table "Proposed Solutions."

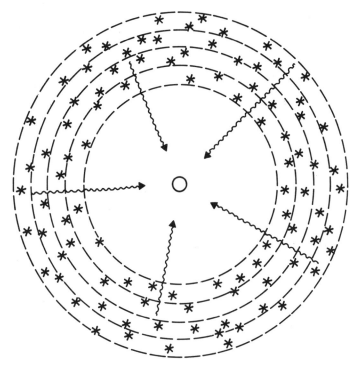

6.5 Large imaginary spheres are constructed with us at the center. These spheres increase in radius and form shells of constant thickness like the layers of an onion. The number of stars in a shell increases with the radius of the shell, but the light received at the center from each star decreases; these two effects (the stars increasing and the light from each decreasing) compensate each other and all shells contribute equal amounts of light.

struct large imaginary spheres of increasing size with the Earth at the center. These spheres form a series of shells of equal thickness, like the layers of an onion. Doubling the size of a shell, keeping its thickness constant, increases its volume fourfold. But the number of stars in any shell is proportional to its volume, and doubling the size of a shell therefore increases fourfold the number of stars it contains.

Doubling the distance of a star reduces the light received from it to one quarter. At the same time, doubling the size of a shell increases the number of stars fourfold; therefore we receive from each shell the same amount of starlight. A distant shell of many faint stars gives as much starlight as a nearer shell of fewer and brighter stars.

The spherical-shell argument is rather general. Stars cluster into galaxies and therefore we should consider a uniform distribution of galaxies. We can increase the size of our imaginary spheres so that the shells accommodate galaxies instead of stars. One shell, double the size of another, then contains four times as many galaxies and hence four times as many stars.

Each shell, according to this argument, contributes a small, fixed amount of light. But a universe of endless space contains an infinite number of shells, implying that we should receive an infinite amount of starlight. Most investigators of the riddle have realized that this conclusion cannot be true. A little thought shows that nearby stars intercept some of the light from stars far away. The stars in each succeeding shell cover the sky slightly more and obstruct our view slightly more of stars farther away. By adding shells, one after another, we ultimately attain a background or cut-off limit when the sky is completely covered with the disks of stars. Very little starlight then reaches us from the more distant shells.

Twenty-four years later the young Swiss astronomer Loys de Chéseaux clarified the confusion in Halley's argument.

A Forest of Stars

From this argument it follows that if starry space is infinite, or only larger than the volume occupied by the Solar System and the first-magnitude stars by the ratio of the cube of 760,000,000,000,000 to 1, each bit of the sky would appear as bright to us as any bit of the Sun, and therefore the amount of light received from each celestial hemisphere—one above and the other below the horizon—would be 91,850 times greater than what we receive from the Sun.

Jean-Philippe Loys de Chéseaux, *Traité de la Comète*

BORN in 1718, the son of a landowner of modest wealth in the Swiss village of Chéseaux near Lausanne, Jean-Philippe Loys de Chéseaux (Figure 7.1) was a precocious and gifted child of delicate health. Educated by his grandfather, the mathematician Jean-Pierre de Crousaz, he developed an interest in astronomy while a youth and constructed his own observatory. At seventeen he wrote papers on the physics of collisions, retardation of cannonballs by air resistance, and sound propagation. Never greatly robust, he died while on a visit to Paris at age thirty-three.

In December 1743, the year following Halley's death, Chéseaux observed a remarkable comet, whose twin tails on the nights of March 7–8, 1744, were seen divided into six tails (Figure 7.2).[1] Only a few months later Chéseaux published a book entitled *Treatise on the Comet of December 1743 to March 1744*. In this work he included a description of the Solar System, an account of how the study of comets benefits astronomy, and a discussion on the computation of cometary orbits. The book concludes with eight appendixes in the form of essays on miscellaneous topics related to astronomy. The second appendix, entitled "On the force of light, its propagation through the ether, and the distance of the fixed stars" (translated and reproduced in Appendix 3), discusses

7.1   Jean-Philippe Loys de Chéseaux (1718–1751).

the dark night-sky riddle and gives for the first time a quantitative analysis of the problem.

Chéseaux made no reference to earlier work on the subject of darkness at night, perhaps because he assumed that readers were familiar with the astronomical literature, particularly with Halley's two short papers published slightly more than twenty years earlier and recently reprinted.[2]

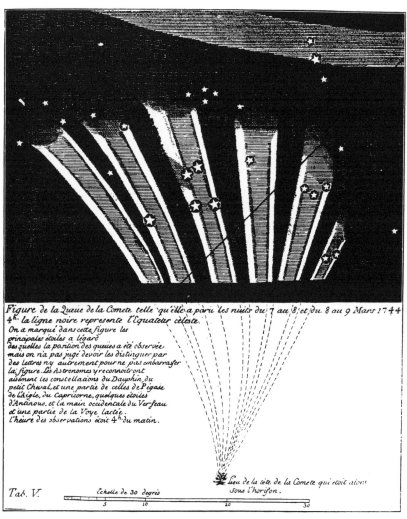

7.2 Chéseaux's comet of 1744 seen at Lausanne and drawn by Loys de Ché-seaux himself. "It appears certain from all the observations up to March 1," he wrote, "that if this comet had appeared under more favorable circumstances, for example in the middle of the night instead of so near the setting Sun, and free of moonlight, it would have been the most striking comet ever known, alike for the size of its head, and the length of its tail, which up to this time had been simply double; but something much more surprising was in store for us." The sky was cloudy until March 7, and then Chéseaux saw with considerable surprise that the fan-shaped tail had in the meantime divided into six branches. After the nights of March 7–8 the comet was not seen again.

Halley had argued that the sky is indeed covered with stars, but that most stars are invisible, despite their infinite number, because their rays are too weak to make an impression on the eye. Chéseaux avoided this error. He followed Halley and imagined, for the sake of computation, that concentric spheres of increasing radius surround the Sun, and the spaces between adjacent spherical surfaces form thin shells of constant thickness, having volumes proportional to their surface areas, and therefore to the square of their radii. If stars are distributed uniformly in space, the number of stars in a shell is proportional to the volume of the shell, hence to the square of the radius. For example, doubling the radius of a shell quadruples the number of contained stars.

Chéseaux assumed that all stars are similar to the Sun. The apparent area of the disk of a star is proportional to the inverse of the square of its distance: doubling the distance reduces its apparent area to a quarter. (He was unaware of the diffraction of light; but as it happens, diffraction is irrelevant in determining the total brightness of the sky; so also is atmospheric blurring and imperfections in optical instruments.) Thus, the increase in number of stars in each successive shell compensates for the decrease in the apparent area of individual stars. All shells, according to this argument, contribute equally to covering the sky with stars, and the observer at the center receives from all shells equal amounts of light.[3]

Distances to the Stars

By observing the Sun in a dark room through a tiny hole in a screen, and by adjusting the amount of light until the image resembled the star Sirius, Christiaan Huygens reckoned that Sirius was 30,000 astronomical units distant. An astronomical unit equals the distance from the Sun to the Earth. This comparison depends on judging in daytime how bright Sirius is at night. Unknown to Huygens, the Scottish astronomer James Gregory had proposed some years earlier a superior method of measuring the distances of bright stars, which did not depend on memory. Gregory assumed that the nearby bright stars are sunlike, and compared their brilliance directly with that of the outer planets Mars, Jupiter, and Saturn. By knowing the sizes of the planets and their distances from the Sun, and allowing for the imperfect reflection of sunlight from their surfaces, he was able to estimate the distances of the brightest stars. Newton referred to Gregory's method in his *System of the World*, and in unpublished work he placed the brightest stars at distances of roughly 500,000 astronomical units; this result, accurate within a factor of two for the nearby stars, remained unknown until 1728.[4]

Light-travel time is a convenient, twentieth-century way of stating large distances in astronomy. Light travels 300,000 kilometers every second, and sunlight takes 500 seconds to reach us from the Sun. One light-second equals 300,000 kilometers, and one light-year equals 10 trillion kilometers, or 63,000 astronomical units. When distances are stated in light-travel time, we realize immediately that the Sun at 500 light-seconds is seen as it was 500 seconds ago. The nearest star, apart from the Sun, is Alpha Centauri (actually three companion stars) at a distance of 4.3 light-years, and we see Alpha Centauri as it was 4.3 years ago.

Applying James Gregory's photometric method, Chéseaux estimated that the brightest stars are 240,000 astronomical units distant, or, in more modern units, 4 light-years. He knew that this was a rough guess because stars have different colors and therefore cannot all be alike. In view of the crudity of the method, his result was remarkably accurate.

Chéseaux's Calculations

By calculation Chéseaux found that a star-covered celestial hemisphere would be 90,000 times brighter than the Sun. This result follows from the assumption that all stars are similar to the Sun, and the fact that the area of the celestial sphere is 180,000 times larger than the apparent area of the Sun.

He then found that 760 trillion shells, each having a thickness of 4 light-years, would cover the entire sky with stars. All these stars would occupy a vast sphere of radius 3,000 trillion (3 followed by 15 zeros) light-years. When the sky is fully covered, no more light can be received by the addition of further shells.

According to Chéseaux's argument, each shell contributes a small and fixed quantity of light. But a universe of infinite extent contains an infinite number of shells. One might therefore conclude that an infinite amount of light should pour down from an infinite number of shells. But Chéseaux avoided this trap. True, each succeeding shell of stars contributes a small quantity of light, and the total amount received increases with the number of shells. When the sky is fully covered, however, no more light can be received by the addition of further shells. The visible stars merge and create a continuous background behind which hide the remaining invisible stars.

Chéseaux's treatment assumed that on the average all places are alike in the universe. Because a system of concentric shells may be constructed about any point in space, always yielding the same result, we conclude that observers at all places in space will see the sky covered with stars.

How far can we see in a universe containing stars that stretch away to unlimited distances? Or, in other words, how far on the average does a line of sight extend into the universe before intercepting the surface of a star? Rather than explain how Chéseaux arrived at his result, on which he himself was not very clear, we shall perform the calculation by a slightly different and more convenient method.

The Forest Analogy

Imagine that we stand in a forest—not a dense jungle nor a tangled scrub but a forest of tall and well-spaced trees (Figure 7.3). The visible trees around us, perhaps as many as several hundred, overlap one another and merge into a continuous background. This background surrounds us like a circular wall; inside are the visible trees of the forest, and outside the invisible trees. If all the trees were visible, and we could look between their trunks and catch glimpses of the world outside, we would hardly call so few trees a forest.

How far can we see in a forest? In some horizontal directions the distance is small and our line of sight intercepts the nearest trees; in other horizontal directions the distance is large and our line of sight probes deep into the forest background. As we gaze around we see an average distance that will be called the *background limit*. This background limit, known in other contexts as the mean free path, is the average distance an arrow in flight travels before striking a tree.

The background limit is easily calculated.[5] Let us suppose the trees are much alike and uniformly distributed; each has a typical width at eye-level and occupies an average area (Figure 7.4). On dividing the average area by the width of a tree trunk, we get the background limit:

$$\text{background limit} = \frac{\text{area occupied by one tree}}{\text{width of tree}}.$$

As an illustration, suppose that the typical separating distance between trees is 10 meters, the average area occupied by one tree is 10 × 10 = 100 square meters. If the trees have a typical width of ½ meter, then the background limit is 100/½ = 200 meters. Of course, some vis-

7.3 "A forest is . . . a certain territory of woody grounds, fruitful pastures, privileged for wild beasts and fowls of forest, chase and warren, wherein to rest and abide, in the safe protection of the king, for his princely delight and pleasure." (Manwood, *Lawes of the Forest*, 1598.)

ible trees are nearer than 200 meters and others farther away, but their average distance equals 200 meters.

The number of trees visible is also easily calculated. This number is very roughly the area of a circle, of radius equal to the background limit, divided by the area occupied by one tree. Using this relation and the equation above, we find:

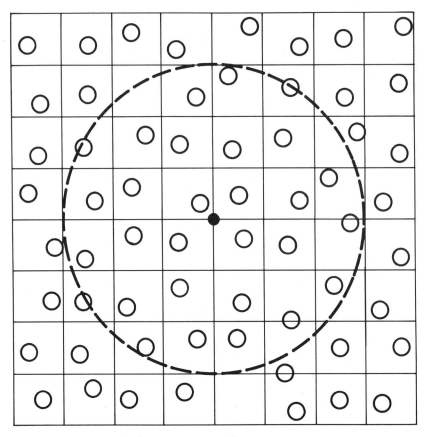

7.4 A diagrammatic forest of uniformly scattered trees. The background limit for an observer at the center of the diagram is the dashed circle. The distance of the background limit from the observer equals the average area occupied by one tree divided by the width of a tree trunk at eye level. In this diagram the average separation is 3 times the width of a trunk, and therefore the background limit is 3 × 3 = 9 times the width of a tree, or 3 times the separation between trees.

$$\text{number of visible trees} = \frac{\pi \times \text{area occupied by one tree}}{(\text{width of tree})^2}$$

where π equals 3.14, or approximately 22/7, and is the ratio of the circumference of a circle to its diameter. In the example of a forest in which each tree has a typical width of ½ meter and occupies an average area of 100 square meters, we find that the number of visible trees amounts to $\pi \times 100/\frac{1}{4} = 1{,}256$. If the separating distance is not 10 but 5 meters, the background limit drops to 50 meters, and the number of visible trees become 314.

This discussion shows us how we can approach the problem of a starry forest. Incidentally, these simple rules for calculating the background limit and number of visible trees in a forest may, with suitable values for the width, be applied to crowds of people and herds of animals.

Optical thickness is a useful technical word. Astronomers refer to anything impenetrable by eye as "optically thick."[6] A cluster of trees extending farther than its background limit, and hence impenetrable by eye, is optically thick. One might say that all forests are, or should be, optically thick by definition. A cluster of trees extending less than its background limit is optically thin. Clumps or small groups of trees are frequently optically thin.

The Starry Forest

The universe is like a starry forest. Yet, very oddly, we fail to see a continuous background of stars. The forest analogy has the merit of revealing immediately what is at issue in the riddle of darkness.

How far can we see in a forest of stars? We suppose that stars are much the same everywhere, and further suppose that they are distributed uniformly throughout illimitable space. Foreground stars, despite their minute geometric size, tend slightly to obstruct our view of more distant stars. In some directions a line of sight intercepts the surface of a nearby star, and in other directions it probes deep into the starry background.

The background limit in an unbounded universe uniformly populated with stars can easily be calculated. A star has a typical cross-section (π times its radius squared) and occupies a typical volume. When the volume is divided by the cross-section, we get the background limit:

$$\text{background limit} = \frac{\text{volume occupied by one star}}{\text{cross-section of star}}$$

The number of stars visible (or rather the number that should be visible) is roughly the volume of a sphere, of radius equal to the background limit, divided by the volume occupied by one star. On using this relation and the equation for the background limit, we find

$$\text{number of visible stars} = \frac{4\pi \times (\text{volume occupied by one star})^2}{3 \times (\text{cross-section of star})^3}$$

This expression gives the number of uniformly scattered stars needed to cover the entire sky. Even in an infinite universe containing an infinite number of stars, we see only a finite distance and a finite number of stars.[7]

These rules for calculating the background limit and the number of visible stars may be applied, with suitable values substituted for the cross-section, to swarms of bees and flocks of birds.

Finding the Background Limit

We are now ready to repeat Chéseaux's calculations with our more convenient equations and up-to-date numbers.

First we notice that within a distance of 10 light-years from the Sun we find about 10 stars. Things will not go far wrong if we assume that in this part of the Galaxy 10 stars occupy a volume of $10 \times 10 \times 10 = 1,000$ cubic light-years, and each star therefore occupies an average volume of 100 cubic light-years.

The Sun's radius is 700,000 kilometers, about twice the distance to the Moon, and therefore its cross-section is 1½ trillion square kilometers. For simplicity we continue to suppose that all stars are similar to the Sun in size and luminosity. If every star occupies an average volume of 100 cubic light-years, as in the neighborhood of the Sun, then the division of this volume by the cross-section shows that the background limit in light-years is 6,000 trillion or 6 followed by 15 zeros. Chéseaux's value for the background limit was one-half this result—less because he used a smaller average volume per star. The number of stars required to cover the whole sky is roughly 10 billion trillion trillion trillion, or 1 followed by 46 zeros.

Perhaps it was the immensity of the background limit, a distance greater than any previously calculated in astronomy, that prompted Chéseaux to think that absorption in space, even the slightest, would veil the most distant stars and create the observed dark night sky.

The Misty Forest

I saw Eternity the other night
Like a great ring of pure and endless light,
All calm, as it was bright,
And round beneath it, Time in hours, days, years,
Driv'n by the spheres
Like a vast shadow moved; in which the world
And all her train were hurled.

Henry Vaughan, *The World*

A FINITE starry cosmos with shores washed by an infinite ocean of emptiness lost its appeal in the eighteenth century. Young Jean-Philippe Loys de Chéseaux, like Edmund Halley in his later years, rejected the Stoic system that had been firmly rooted in the seventeenth and earlier centuries; he much preferred the Cartesian and Newtonian endless forest of stars. Yet theory showed that in an infinite, star-filled universe the sky should blaze at every point with starlight. In his essay "On the force of light, its propagation through the ether, and the distance of the fixed stars" he wrote,

> The enormous difference between this conclusion and experience demonstrates either that the sphere of fixed stars is not infinite but actually incomparably smaller than the finite extension I have supposed for it, or that the force of light decreases faster than the inverse square of distance. This latter supposition is quite plausible, it requires only that starry space is filled with a fluid capable of intercepting light very slightly.

Chéseaux thus solved, or thought he solved, the riddle of the dark night sky by assuming that starlight is slowly absorbed while traveling across the immense gulfs of interstellar space.*

Chéseaux was a Cartesian to the extent of assuming automatically

*Entry 6 in the table "Proposed Solutions."

that a material medium pervades interstellar space. And this medium, though highly rarified, need not be perfectly transparent and might therefore slowly attenuate light as it travels from the stars. Thus, a fog fills interstellar space and obscures from view all the remoter stars, in the same way that a mist obscures from view the background trees in a forest.

Chéseaux's solution requires that rays of starlight are absorbed in a distance much less than the background limit.[1] If space is not perfectly transparent, said Chéseaux, but 330,000 trillion times more transparent than water, the total amount of starlight incident on Earth would be only 33 times brighter than earthshine on the new Moon. (Earthshine is the reflection of the Earth's light.) This amount of starlight could easily be brought into agreement with observation by considering a slightly different value for the transparency of interstellar space. To pursue Chéseaux's solution, grasp its significance, and understand why it fails, we must jump ahead, first to the work of Wilhelm Olbers in the nineteenth century.

Wilhelm Olbers

Born in 1758 at Arbergen, Olbers (Figure 8.1) practiced in Bremen as a successful physician, specializing in ophthalmology (then in its infancy), advocating innoculations, and receiving praise for his work in combating cholera epidemics. From early youth, astronomy fascinated him, and he gained international fame for his studies of comets and the discovery of the asteroids Pallas and Vesta. Sleeping only four hours in twenty-four, he lived by day as a busy physician and by night as a diligent astronomer. In 1820, following the death of his daughter and second wife, he retired from medical practice and devoted his remaining years to astronomy.

In 1823 Olbers published an important scientific article bearing the title "On the transparency of space" (translated in Appendix 4).[2] He wrote, "Is space not infinite? Are limits to it conceivable? And is it conceivable that the omnipotence of the Creator would have left this interminable space empty?" He quoted from Kant's work on the infinity of space and then said, "Quite probably not only the portion of space that our eye has penetrated with the aid of instruments, or may be penetrated in the future, but also endless space itself is sprinkled with suns and their accompanying planets and comets." He cautioned the reader not to forget that other regions of space may contain creations different

8.1 Heinrich Wilhelm Matthaus Olbers (1758–1840), a man of "successful practice and integrity and affability of character." (Thomas Dick, *Celestial Scenery, or the Wonders of the Planetary System Displayed*, 1838, p. 127n.)

from the suns, planets, and comets seen in our heavens and unlike anything we can imagine.

Olbers referred to Halley's work and pointed out that Halley had failed to show clearly why the sky at night is dark in a star-filled universe of infinite extent. The problem of an infinite universe raised by Halley could not be dismissed as easily as he had supposed. Olbers made the interesting remark,

> If there really are suns throughout the whole of infinite space, and if they are placed at equal distances from one another, or grouped into systems like that of the Milky Way, their number must be infinite and the whole vault of heaven will appear as bright as the Sun; for every line that we can imagine drawn from our eyes would necessarily lead to some fixed star, and therefore starlight, which is the same as sunlight, would reach us from every point of the sky.

Here we encounter for the first time the realization that stars need not be uniformly distributed and may be grouped into milky way systems that nowadays we call galaxies. "Clearly," said Olbers, "whether stars are uniformly distributed in space or clustered in widely separated systems, the same conclusion follows." Also for the first time we encounter the powerful line-of-sight argument. A line extended from the eye in any direction eventually terminates at a point on the surface of a star, and this occurs regardless of whether stars are distributed uniformly or clustered into galaxies. A ray of starlight follows the line of sight in reverse direction and eventually enters the eye. "Not only the whole celestial vault will be covered with stars," he said, "but also stars will stand one behind another in endless rows covering each other from our view."

In a forest that extends farther than its background limit, we always observe an impenetrable background, however widely spaced or clustered the trees or however narrow their trunks. Similarly, in a universe that extends farther than its background limit, we expect always to observe an impenetrable background of stars, however widely spaced or clustered the stars or however small their size. Metaphorically, we stand in a forest composed of clumps of trees, in which a tree represents a star and a clump of trees represents a galaxy (Figure 8.2). Even when we can see through a clump of trees, the superposition of many clumps still creates a continuous background of trees. Similarly, in a universe of galaxies, even when we can see through a galaxy of stars, the superposition of many galaxies still creates a continuous background of stars. Wherever we stand in a universe of galaxies, an impenetrable back-

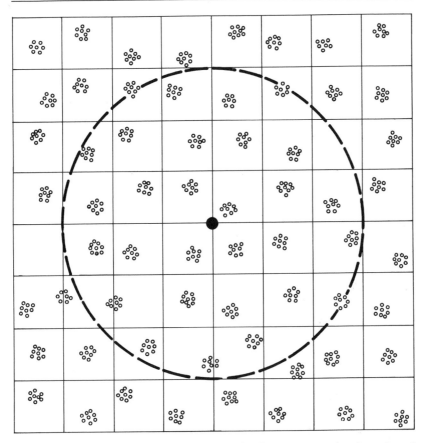

8.2 Olbers introduced the important line-of-sight argument. The clustering of stars (trees) does not affect the argument that every line of sight intercepts the surface of a star (tree). This is obviously true when trees cluster together in dense clumps that cannot be seen through, as in this illustration. The background limit equals the average area occupied by one clump divided by the width of a clump. It is also true when trees cluster together in loose clumps that can be seen through. The background limit then equals the average area occupied by one tree divided by the width of a tree, and clumping in this case has little effect on the background limit.

ground of stars exists, or should exist, because every line of sight terminates at the surface of a star.

A Strange Coincidence

Olbers, like Chéseaux, used Gregory's photometric method of determining stellar distances and found that first-magnitude stars are at a distance of about 350,000 astronomical units, or 5½ light-years. This value came close to that obtained by Newton and Chéseaux. Optical instruments had improved to the stage where soon, in 1838, more precise methods of measuring distance would become possible. Friedrich Bessel, Olbers's protégé, succeeded in measuring for the first time the distance of a nearby star by means of parallax.

Like Chéseaux, Olbers constructed concentric spherical surfaces of increasing radius, creating in this way a series of imaginary shells of the same thickness. He then, in effect, summed the light from all shells out to the background limit. But his firm conviction that the whole sky is covered with stars caused him no anxiety because he knew, or thought he knew the correct answer to the riddle of darkness. "Because the celestial vault has not at all points the brightness of the Sun, must we reject the infinity of the stellar system? Must we restrict this system of stars to one small portion of limitless space?"

> Not at all. In our inferences drawn from the hypothesis that an infinite number of fixed stars exists we have assumed that space throughout the whole universe is absolutely transparent, and that light, consisting of parallel rays, remains unimpaired as it propagates great distances from luminous bodies. Now this absolute transparency of space is not only undemonstrated but also highly improbable.

His proposed solution—the absorption of starlight in interstellar space—was the same as that already given by Chéseaux almost eighty years previously (Figure 8.3).

Olbers possessed in his library a copy of Chéseaux's book on the comet of 1744.[3] Yet, unaccountably, he failed to acknowledge this earlier work that anticipated and closely resembled his own. He may have assumed that astronomers were aware of Chéseaux's book and therefore supposed that acknowledgment was unnecessary. Much more probably, he forgot that he had read Chéseaux's book years before and consequently supposed that he himself had originated the idea of absorption. I am inclined to think that claiming an idea and forgetting its source occurs more often than generally accepted. Owing to Olbers's

8.3 A misty forest in which background trees are obscured from view and only foreground trees can be seen.

negligence—the most we may call it—the riddle in recent times became known not as Chéseaux's paradox but as Olbers's paradox.

A White-Hot Furnace

Neither Chéseaux nor Olbers dreamed of the conservation of energy, or even thought of heat and light as convertible forms of energy, and neither could therefore appreciate the full implication of a bright-sky universe. A bright sky pouring down 90,000 times as much light as the Sun did not seem to them to be so terribly disastrous. Olbers even said that life might adapt to the continuous glare of a sky fully covered with stars. His main concern was the plight of astronomers: "Astronomy for the inhabitants of the Earth would remain forever in a primitive state; nothing would be known about the fixed stars; only with difficulty would the Sun be detected by virtue of its spots; and the Moon and planets would be distinguished merely as darker disks against a background as brilliant as the Sun's disk."

In Chéseaux's day, and even later in Olbers's day, astronomers were unable to appreciate that the conditions in a bright-sky universe would be much like those in a high-temperature furnace. Nowadays, with little effort on our part, we understand that the Sun is bright because its surface has a temperature of 6,000 degrees Kelvin. (Degrees Kelvin are measured from absolute zero at minus 273 degrees Celsius. Water freezes at 273 degrees and boils at 373 degrees Kelvin.) The Sun's hot surface radiates into space an immense amount of energy over a wide spectrum of wavelengths, visible and invisible to the eye, and this radiant energy, when incident on matter and absorbed, transforms back into thermal energy, or heat.

Imagine the sky ablaze at every point as bright as the Sun's disk—as if the Earth had plunged into the photosphere of the Sun. What would the conditions be like? They would be much the same as if the Earth were inside a furnace, surrounded by white-hot walls at the temperature of the Sun's surface. William Herschel, the foremost astronomer of the eighteenth century, whose work we shall soon meet, thought that perhaps living creatures dwelt on the surface of the Sun. Until the development of thermodynamics, temperature was not a very clear concept, and like Olbers, Herschel did not associate the brightness of the Sun's surface with a temperature too high for life.

Let some object, say a metal ball, be placed in a vacuum furnace. The ball heats up and becomes as hot as the inside of the furnace. At a sufficiently high temperature the ball melts and turns into a liquid, and at even higher temperature it evaporates and turns into a gas. We now introduce an inert atmosphere such as argon into the furnace and repeat the experiment. The inert gas heats up and the results are the same as before. The gas only slightly delays the heating process. Next we enclose the ball in a refractory insulating material and again repeat the experiment. The insulating material heats up, then radiates as much energy as it asborbs, and the results are again much the same as before. The absorbing medium, while heating up, provides only temporary insulation.

Thus, Chéseaux's and Olbers's idea of transforming a bright-sky universe into a dark-sky universe by means of interstellar absorption is rather hopeless. The absorbing medium soon heats up and reaches the temperature characteristic of the surfaces of stars. The medium then radiates as much energy as it absorbs, and we are no better off than before.

John Herschel—a famous scientist and son of the astronomer William Herschel—commented on Olbers's idea in *The Edinburgh Review* in 1848 when he reviewed Alexander von Humboldt's *Kosmos*. "The assumption that the extent of the starry firmament is literally infinite has been made by one of the greatest astronomers, the late Dr Olbers, the basis of a conclusion that the celestial spaces are in some slight degree deficient in *transparency;* so that all beyond a certain distance is, and must forever remain, unseen . . . Were it not so, it is argued, every part of the celestial concave ought to shine with the brightness of the solar disc."[4] He then explained why absorption could not solve the riddle of darkness:

Light, it is true, is easily disposed of. Once absorbed, it is extinct forever and will trouble us no more. But with radiant heat the case is otherwise. This, though absorbed, remains still effective in heating the absorbing medium, which must either increase in temperature, the process continuing, *ad infinitum,* or, in its turn becoming radiant, give out from every point at every instant as much heat as it receives.

John Herschel's comments, made before the law of conservation of energy was established and fully understood, seem remarkable. Though he failed to realize that light is an interchangeable form of energy, and his comments reveal the confusion existing in thermodynamics in his day, he at least understood that radiant heat, when absorbed, creates heat in material bodies that again is emitted as radiant heat. He saw that the continual absorption by the interstellar medium of the radiant heat emitted by stars, as in the solution proposed by Chéseaux and Olbers, would cause the absorbing medium to heat up and eventually radiate "from every point at every instant as much heat as it receives."

Hermann Bondi, in his book *Cosmology* (1960), which in recent years reawakened interest in the riddle of cosmic darkness, says much the same: "What happens to the energy absorbed by the gas? It clearly must heat the gas until it reaches such a temperature that it radiates as much as it receives, and hence it will not reduce the average density of radiation."[5] We may add that the temperature of the gas at which this occurs equals the average temperature of the surfaces of stars.

But claims that absorption fails to prevent a catastrophic bright sky must themselves be critically examined. If the average lifetime of a star is 10 billion years and the absorbing medium takes only 1,000 years to heat up, absorption is completely ineffective, as Herschel claimed. Absorption is effective, however, when the absorbing medium heats up in a period of time longer than the lifetime of luminous stars.

Herschel's remarks do not apply to the solution suggested by Edward Fournier d'Albe. This British scientist-engineer devoted most of his life to developing science in Ireland and promoting anglo-celtic relations. In 1923 he successfully transmitted television pictures from London. Fournier d'Albe haunts our story; he flits through the pages of *Darkness at Night* like a ghost wandering from corridor to corridor at midnight, and we shall meet him again and again. His short book *Two New Worlds*,[6] published in 1907, is a mine of cosmological speculation containing many gems. He stated the riddle of cosmic darkness in the words: "It has been shown with mathematical certainty that if infinite space were strewn with stars, as is our stellar system, and if these stars shone like our stars, and if they had existed for all eternity, then the appearance of the sky would be one blaze of sunlight throughout." He stressed the consequences of this dreadful state of affairs and showed that absorption offers no relief:

> To this I may add the equally certain conclusion that every part of space, including the earth and ourselves, would be at a white heat and gaseous; for radiant heat would be propagated in the same way as light. And, further, that the circumstance of space being cold can be used as a conclusive proof of the absence of any sensible loss of radiation through absorption in space; for the absorbing medium would be heated by the process of absorption, and would then itself radiate the heat inwards upon our devoted heads.

Fournier d'Albe considered several solutions, but the one of immediate interest to us depended on the novel idea that most stars are normally nonluminous: "If a hot star is something altogether exceptional—a freak happening in a billion times—then the average temperature of an infinite universe will be quite comfortable." His billion—a million million—equals an American trillion. A trillion nonluminous stars to every luminous star converts a bright-sky universe into a dark-sky universe and overcomes Herschel's objection.* In normal circumstances non-luminous stars will stay relatively cold and will not heat up in a luminous lifetime. Thus, if x is the fraction of luminous stars (1 part in a trillion in Fournier d'Albe's example), then x is also the fraction of the sky covered with luminous stars. In this solution dark stars fill the dark gaps between visible stars. Modern knowledge of the composition of the interstellar medium and structure of the universe rules out such an extreme solution, however, and we may safely say that absorption of starlight is not the explanation of a dark night sky.

Having ruled out absorption, we now return to the eighteenth century and pick up the threads of our tale. The one-island Stoic system has been set aside, although only temporarily, and the unending space populated with endless stars of the Epicurean system has been restored in the waning Cartesian and waxing Newtonian universes. Yet dissenting voices can be heard proclaiming the possibility of a grander universe composed of numerous Stoic systems—a many-island universe.

*Entry 7 in the table "Proposed Solutions."

NINE

Worlds on Worlds

He, who through vast immensity can pierce,
See worlds on worlds compose one universe,
Observe how system into system runs,
What other planets circle other suns,
What varied Being peoples every star,
May tell why Heaven has made us as we are.

Alexander Pope, *Essay on Man*

ACROSS the night sky arches the wraithlike Milky Way. Throughout recorded history astronomers have pondered on the nature of this band of encircling nebulous light. Democritus said the Milky Way consists of shining stars, too numerous and too distant to be resolved by the eye, and Galileo said much the same. Isaac Newton, in his unpublished *Cosmography,* wrote, "For ye milky way being viewed through a good Telescope appears very full of very small fixt stars & is nothing else then ye confused light of these stars." We cannot avoid glancing at this fascinating subject of groping and emerging ideas about the Milky Way, our Galaxy, and about other milky ways, or galaxies, for its history intertwines with the history of the riddle of cosmic darkness.

In the twentieth century, with the benefit of greater knowledge, we look out from the Solar System to the star throngs and gas clouds that sprawl across the sky and constitute the whirlpool system of our Galaxy. Starlight from the center of the Galaxy takes 100 thousand years to escape across the disk into the depths of intergalactic space, and two million years to reach the neighboring giant spiral galaxy in Andromeda, which is similar to our Galaxy.

We look out from the Galaxy, beyond the Local Group of twenty or so galaxies, and beyond the swarms of neighboring groups of galaxies, to the giant clusters of hundreds and sometimes thousands of galaxies. We look out far beyond the Local Supercluster to myriads of superclusters—each hundreds of millions of light years across—scattered to the horizon of the visible universe at a distance of approximately 15 billion light-years.

Thomas Wright's Speculations

Born in 1711, Thomas Wright of Durham in Northern England, at first an unsettled youth who ran away to sea, then later in life an estate surveyor, author, science and mathematics teacher, must be given credit for the idea that the Milky Way consists of a layer of stars.[1] The Sun is one of the many stars in this layer, he said, and from our position, looking out in directions more or less parallel with the layer, we see light from multitudes of stars coalescing to form a milky appearance, and in directions more or less perpendicular to the layer, we see between the stars the darkness of outer space (Figure 9.1).

Wright advanced this germinal idea and other thoughts in 1750 in his last major work entitled *An Original Theory or New Hypothesis of the Universe.*[2] In this splendidly illustrated book, published a year before the death of Loys de Chéseaux, he proposed:

(1) The Milky Way consists of a layer of stars that includes the Sun, and this layer exists in the form of either a disk or a spherical shell (Figure 9.2).

(2) The stars are not fixed but in motion, as first noticed by Halley, and they orbit around the center of the Milky Way like planets around the Sun. All the moving stars of the Milky Way respond to "centrifugal Force, which not only preserves them in their Orbits, but prevents them from rushing all together, by the common universal Law of Gravity."

(3) Perhaps other milky ways exist at vast distances. "That this in all Probability may be the real Case, is in some Degree made evident by the many cloudy Spots, just perceivable by us, as far without our starry regions, in which . . . no one Star or particular constituent Body can possibly be distinguished."

Each of these original proposals by this extraordinary man have since borne fruit. The third proposal—the inspired notion that the small fuzzy patches of light seen in the heavens are perhaps other milky ways—opened up a truly spectacular view of the universe (Figure 9.3).

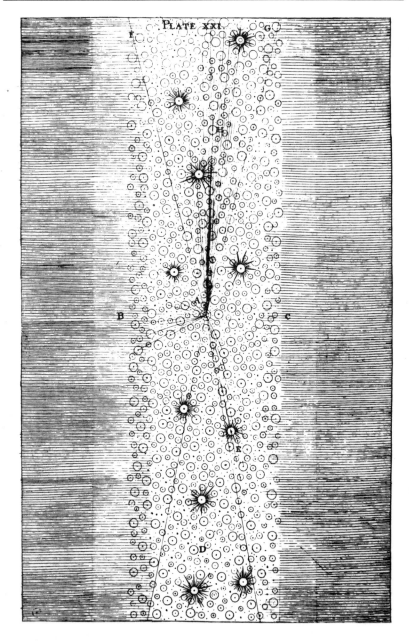

9.1 Thomas Wright of Durham proposed the original idea that the Milky Way consists of a layer of stars, as shown in his beautifully illustrated book *An Original Theory or New Hypothesis of the Universe* (1750).

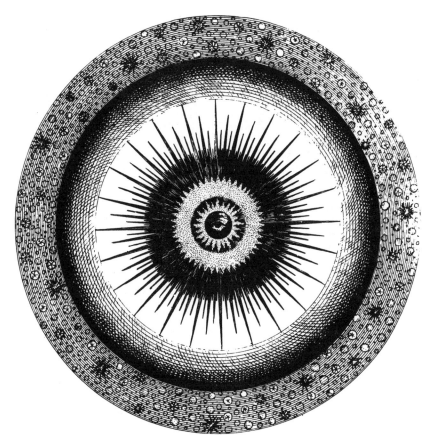

9.2 Alternatively, said Wright, the stars occupy a shell around a mysterious galactic center, as shown in this illustration.

The Cloudy Stars

Unavoidably our story of "worlds on worlds" enters a period in which two complex multistranded themes—the story of the galaxies and the story of the nebulae—intertwine, often in a most confusing manner. The nebulae, those faint and fuzzy patches of light in the heavens, a few of which Ptolemy recorded, and often called "cloudy stars" by astronomers, began to attract serious attention after the discovery of the telescope. Galileo thought they were concentrations of stars and wrote in *The Starry Message* that "the galaxy" consists of nothing but "innumerable stars grouped together in clusters."

9.3   Wright also proposed the existence of distant milky ways (galaxies) similar to our Milky Way (Galaxy). In this illustration in *An Original Theory or New Hypothesis of the Universe* he showed the distant milky ways covering the whole sky.

But it is not only in the Milky Way that whitish clouds are seen; several patches of similar aspect shine with faint light here and there throughout the aether, and if the telescope is turned upon any of them it confronts us with a tight mass of stars. And what is even more remarkable, the stars which have been called "nebulous" by every astronomer up to this time turn out to be groups of very small stars arranged in a wonderful manner.[3]

Galileo then said, and as usual, his comments strike a profound note: "Although each star separately escapes our sight on account of its smallness or immense distance from us, the mingling of their rays gives rise to that gleam" that formerly was believed to be the reflection of sunlight or starlight off denser regions of the ether. He rejected this Aristotelian explanation of nebulae, and furthermore, he understood what Halley failed to appreciate: that even though individual stars may escape our sight, many stars seen together can create a visible gleam.

But astronomers continued to find that no matter how powerful their telescopes, not all nebulae could be resolved into stars. Indeed, Pierre Maupertuis, an eighteenth-century French mathematician of wide interests, believed that the fuzzy, often elliptically shaped patches of light in the sky, usually less bright and always more extended than normal stars, were individual abnormal stars, flattened by rotation.[4] Inspired by the ideas of Wright and Maupertuis, Immanuel Kant suggested that some nebulae in fact might be new stars or solar systems condensing out of gas, and yet other nebulae might be whole stellar systems—milky ways similar to our Milky Way—so immensely distant that we see them as dim patches of light.

Kant and the Evolving Universe

Kant was born in 1724 in Königsberg (the capital of Prussia, now in the Soviet Union, and renamed Kaliningrad), where he died eighty years later. Although he never traveled away from his hometown, in the words of Lucretius referring to Epicurus, "he ventured far beyond the flaming ramparts of the world and voyaged in mind throughout infinity."[5] At first his main interests were in mathematics and natural sciences. After years of lecturing he was appointed professor of logic and metaphysics at the University of Königsberg. By that time his interests had turned to philosophy, and in 1781 he published the famous *Critique of Pure Reason*, devoted to establishing a theory of knowledge.

Kant saw only a summary of Wright's book in a review article that appeared in a Hamburg journal in 1751.[6] From it he learned that the stars of the Milky Way were perhaps attracted toward a common center, about which they moved in much the same way as the planets of the Solar System. This idea, proposed—according to the report—by the ingenious English author Thomas Wright, conjured up in Kant's mind a picture of a rotating Galaxy consisting of a disk-shaped system of stars.

In 1755, the year he obtained his doctorate, Kant wrote *Universal Natural History and Theory of the Heavens,* a landmark in the history of astronomy. Unfortunately, the publisher went bankrupt, and few copies at the time came into the hands of the public. The work nonetheless caused a stir in scientific circles through the circulation of excerpts and the appearance of a summary in the preface of another book by Kant, *The Only Possible Proof of the Existence of God.*

Paying tribute to Wright, Kant wrote that Wright "regarded the Fixed Stars not as a mere swarm scattered without order and without design," but as a systematic arrangement filling the whole of space. Wright's first two conjectures that the Milky Way consists of stars orbiting in a plane explained why the eye, when located in this plane, sees "on the spherical cavity of the firmament the densest accumulation of stars in the direction of such a plane," and sees this band of stars as "a uniformly whitish glimmer—in a word, a Milky Way."

In the beginning, said Kant, the universe—created by God—was a chaos of atoms governed by natural laws. Out of the vorticity of atom streams, under the urge of gravity, the planets, stars, and galaxies formed. Star formation and other formation processes still continue and perhaps will continue throughout eternity. "Millions and whole myriads of millions of centuries will flow on, during which always new worlds and systems of worlds will be formed . . . The creation is never finished or complete."

Pierre Laplace, whose brilliance as a mathematician earned him the title "Newton of France," discussed in his popular *System of the World* the idea that the Solar System, and other solar systems, originated from contracting and rotating clouds of gas. Kant had earlier advanced much the same thought; Laplace, however, stressed the importance of rotation in the formation of planetary systems, and earns most of the credit. The Kant–Laplace nebula hypothesis, despite criticisms and opposition, caught the imagination of astronomers in the nineteenth century (Figure 9.4) and, with many improvements, forms the basis of our modern theories on the formation of the Solar System.

Astronomers from the time of Galileo have recognized many nebulae as star clusters in which each star "escapes our sight on account of its immense distance." But since the eighteenth century astronomers have

9.4 William Parsons (1800–1867), an Irish astronomer and third Earl of Rosse, constructed giant reflecting telescopes and discovered that some nebulae have spiral structure. This sketch in 1845 depicts the spiral nebula M 51 as seen in his 72-inch telescope. The spiral patterns were attributed to swirling gas clouds in conformity with the Kant–Laplace nebula hypothesis.

also recognized the possibility that many other nebulae may be swirling clouds of gas contracting to form new solar systems, or distant milky ways, later referred to as "island universes." Kant took the liberal, and as it happens the correct, view that many nebulae are loose star clusters, some are gas clouds, and yet others are in fact milky ways. We shall return to the imaginative Kant in a discussion of the fractal universe.

The theories of Thomas Wright and Immanuel Kant far outweighed the discoveries made by astronomical observations. The time was ripe for William Herschel, a skillful maker of telescopes and a diligent observer of the heavens, to restore the balance between theory and observation.

On the Construction of the Heavens

Wilhelm Herschel was born in Hanover, Germany, but emigrated to England in 1757 at age nineteen to escape the hardships of campaigning in the Seven Years War. A musician by profession, he became interested in astronomy, and during his leisure time he constructed telescopes and observed the heavens. In 1772 his young sister Caroline joined him. Both followed musical careers and shared a keen interest in astronomy. Using reflecting telescopes which they made themselves, William and Caroline succeeded in resolving numerous nebulae into stars. Eventually, with Caroline's assistance, William became recognized as the foremost astronomer of the eighteenth century.

In 1781 he discovered a previously unidentified seventh planet—later known as Uranus—beyond the orbit of Saturn, and called the newfound planet Georgium Sidus in honor of King George III.[7] The king appointed William court astronomer with a modest stipend in recognition of this great discovery. In 1788 William married a wealthy widow, and he and his sister abandoned their musical careers and became full-time astronomers.

Brother and sister surveyed the heavens with telescopes of unrivaled precision and light-gathering power. William interpreted their results by arguing as follows: first, space is totally transparent, hence stars are unobscured; second, all stars are similar to the Sun, hence the farthest shine the weakest; third, stars are scattered evenly, hence the finite Milky Way stretches the farthest in those directions where the sky seems the most crowded with stars. Aided by these assumptions and arguments, he and Caroline charted the Milky Way and found it to be a flattened system with the Sun near the center (Figure 9.5).

William thought that certain dark and starless regions in the sky, nowadays known to be dark clouds of obscuring dust, are "holes in the sky," and that through these holes in the Milky Way we see the darkness of extragalactic space. He was unaware of the pronounced absorption of starlight by dust in the plane of the Milky Way and had no idea that the center of our Galaxy in the constellation of Sagittarius is hidden from view. He concluded that we occupy a central position because the Milky Way encircles us with almost uniform brightness.

But when William conducted his pioneering studies of binary systems (two stars in orbit about each other), he realized that his second assumption—all stars are alike—was untenable. So also was his third assumption, for the stars evidently were scattered with pronounced ir-

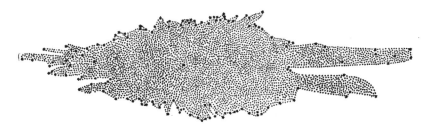

9.5 William Herschel's model of the Milky Way, with the Sun located at the center. This "cloven hoof" side view shows the Galaxy extending the greatest distance where the Milky Way appears the most crowded with stars.

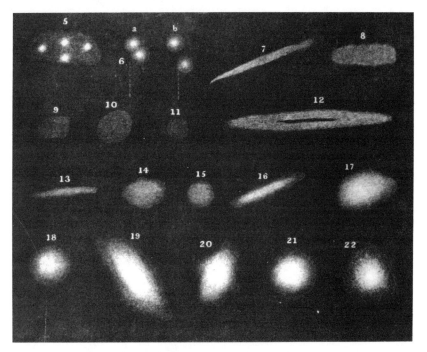

9.6 Part of William Herschel's sketch of various diffused nebulosities in his paper read before the Royal Society in June, 1811: "Astronomical observations relating to the construction of the heavens, arranged for the purpose of a critical examination, the result of which appears to throw some new light upon the organization of the celestial bodies."

regularity. Moreover, some nebulous regions, as in Orion, looked like "a shining fluid of a nature unknown to us," and appeared to consist of more than just unresolved stars, indicating that the first assumption was on shaky ground.

In 1785, in an article "On the construction of the heavens," William announced that many nebulae may be distant systems similar to our Milky Way. "For which reason they may also be called milky ways by way of distinction."[8] A year later he wrote in a letter that he had "discovered 1500 universes! . . . whole sidereal systems, some of which might well outvie our Milky Way in grandeur."[9] For some years his ideas resembled Wright's ideas, and in cautious language he supported the theory of an unending universe populated with endless galaxies, each like the Milky Way. Late in life, however, he changed his mind and expressed the opinion that probably most nebulae (Figure 9.6) lie inside the Milky Way, and external milky ways, if they exist, are much too distant and faint to be detected. He lost his former confidence, renounced the many-island universe of Wright and Kant, and adopted a one-island universe. Following his lead, the one-island universe was widely adopted in astronomical circles in the nineteenth century.

Revelations of Chaos

Lady Constance, after reading *The Revelations of Chaos:*

"It shows you exactly how a star is formed; nothing could be so pretty. A cluster of vapour—the cream of the Milky Way; a sort of celestial cheese churned into light."

William Huggins, "The New Astronomy"

★

WITH few exceptions, astronomers in the nineteenth century had little doubt that our Galaxy encompassed the whole of material creation. Beyond the Galaxy stretched a mysterious infinite space in which the existence of other galaxies or milky ways was no more than a fanciful speculation.

John Herschel, born in 1792 and the only child of William, first studied law and then, disillusioned, turned to science and astronomy. In 1834 he set up an observatory at Cape Colony, South Africa, and mapped the southern skies using much the same principles his father had used for the northern skies. In his famous textbook, *Outlines of Astronomy* (1849), he doubted the existence of any "physical distinction between nebulae and clusters of stars."[1] Observations reveal, he said, that all nebulae are clusters of stars either inside or on the outskirts of our Galaxy, and the Galaxy is the largest object in the universe. If the faint nebulae in the sky away from the plane of the Milky Way are indeed external systems, why are they seen only when we look away from the plane of the Milky Way on either side? Their peculiar distribution surely proves that they belong to our sidereal system. He rejected the explanation, which astronomers now accept, that light traveling in the plane of the Milky Way and coming from external systems is extinguished by absorption.

John Herschel was convinced that the Galaxy is an all-encompassing sidereal system and that we look out through the gaps between the stars and see the blackness of infinite extragalactic space. Most astronomers of the nineteenth century shared this Stoic view. Agnes Clerke (1842–1907), an English astronomer and historian of science, summed up the prevailing belief in 1890 in her *System of the Stars:*

> The question of whether the nebulae are external galaxies hardly any longer needs discussion. It has been answered by the progress of discovery. No competent thinker, with the whole of the available evidence before him, can now, it is safe to say, maintain any single nebula to be a star system of co-ordinate rank with the Milky Way. A practical certainty has been attained that the entire contents, stellar and nebular, of the sphere belong to one mighty aggregation, and stand in ordered mutual relations within the limits of one all-embracing scheme—all-embracing, that is to say, so far as our capacities of knowledge extend. With the infinite possibilities beyond, science has no concern.[2]

Wilhelm Olbers did not accept the one-island universe promoted by the Herschels. In a certain sense, neither did Friedrich Struve, the first in a dynasty of famous Struve astronomers, who fled from Germany to Russia to escape conscription in the armies of Napoleon and became the first director of the Russian observatory at Pulkova, south of St. Petersburg (now Leningrad). He argued strongly that interstellar space cannot be fully transparent. We see into the far depths of starless space when looking away from the Milky Way, he said, but see only a limited distance when looking in the plane of the Milky Way. Struve received considerable criticism for proposing more or less the modern view.[3]

Struve's solution to the riddle of darkness was a compromise: away from the plane of the Milky Way the sky is dark because of the emptiness of extragalactic space; in the plane of the Milky Way, which he thought extended an unlimited distance, the sky is moderately dark because of the absorption of starlight.

The New Astronomy

In 1897—the diamond jubilee year of Queen Victoria's reign—the eminent astronomer William Huggins wrote an essay, "The new astronomy: a personal retrospect," in which he recalled his experiences. As a young man, he wrote, "I soon became a little dissatisfied with the routine character of ordinary astronomical work, and in a vague way sought . . . for the possibility of research upon the heavens in a new direction or by new methods."[4] News of Gustav Kirchhoff's discoveries in spectrum analysis in Germany "was to me like the coming upon a spring of water in a dry and thirsty land." The French philosopher Auguste Comte, inventor of the word "sociology," had declared in his *Positive Philosopy* that astronomers would never be able to determine the composition of stars.[5] But Kirchhoff had shown that the same elements existed in the Sun as on Earth, and Huggins, undeterred by Comte's circumscription, resolved to do the same for the stars.

For several years Huggins pioneered novel techniques in spectroscopy and photography adapted to the observatory. He calibrated the spectra of elements against the spark-gap spectrum of air, and then studied the spectral composition of light from stars, nebulae, and comets. In 1863 he announced the amazing news that, according to his observations, stars consist of elements identical to those on Earth and in the Sun. Working in his home-observatory at Tulse Hill outside London, this unassuming man, who nowadays would be regarded as an amateur, finally overthrew the ancient belief that the heavens are composed of ethereal substances not to be found on Earth.

A year later, "On the evening of the 29th of August, 1864, I directed the telescope for the first time to a planetary nebula in Draco. The reader may now be able to picture to himself to some extent the feeling of excited suspense, mingled with a degree of awe, with which, after a few moments of hesitation, I put my eye to the spectroscope. Was I not about to look into a secret place of creation?" But the spectrum was not what he expected. "A single line only!" He wondered if the spectroscope needed readjusting, "then the true interpretation flashed upon me." This particular nebula lacked the continuous spectrum emitted by stars, and had a spectrum characteristic of a hot gas. "The riddle of the nebulae was solved. The answer, which had come to us in the light itself, read: Not an aggregation of stars, but a luminous gas."

Huggins's discovery, showing that at least some unresolved nebulae consist of luminous gas, overcame all lingering opposition to the nebula hypothesis proposed by Kant and Laplace. According to this theory the Solar System and other planetary systems condensed out of swirling and contracting gas clouds. But, alas, Huggins went too far. The elliptical and spiral-shaped nebulae, he said, such as the great spiral in Andromeda (Figure 10.1), are "not clusters of suns, but gaseous nebulae which, by the gradual loss of heat, or the influence of other forces, have become crowded with more condensed and opaque portions."[6] This sweeping interpretation caused the theory of a many-galaxy universe, proposed by Wright and Kant, to reach the extreme of unpopularity.

In 1868 Huggins successfully measured the radial velocity of stars, moving either away from or toward us, by observing the shift in their

10.1   The Andromeda Nebula, M 31, a large neighboring spiral galaxy, similar to our own Galaxy and at a distance of 2 million light-years. (Courtesy of Kitt Peak National Observatory, Cerro Tololo Inter-American Observatory.)

spectral lines that Armand Fizeau had predicted in 1848. Shortly before, in 1843, Christian Doppler had suggested that stars in binary systems periodically change their color, becoming blue when approaching and red when receding.[7] The displacement in the lines of a spectrum caused by relative motion was known in the nineteenth century as the Fizeau-Doppler effect and is known today, less justifiably, as the Doppler effect. Detection of the Doppler effect was difficult until photography advanced to the stage where dry and sensitive plates could be exposed for hours in clock-driven telescopes. On the subject of the Doppler effect, Huggins prophetically wrote, "It would be scarcely possible . . . to attempt to sketch out even in broad outline the many glorious achievements which doubtless lie before this method of research in the immediate future."[8]

An unsigned article in a magazine on how to make your own spectroscope launched Margaret Lindsay into the new science of spectroscopy.[9] She was a keen photographer and also an observer of the heavens with instruments made by herself. By chance she met the author of the article—William Huggins—who was visiting her home city, Dublin, to inspect his new telescope manufactured by Howard Grubb. Spectroscopy sparked romance, and in 1875 they married. Working as a team— he with failing eyes and she keen-eyed—they made observations using state-of-the-art photography and spectroscopy that helped to launch the new science of astrophysics (Figure 10.2). Queen Victoria in 1897 knighted William by conferring the Order of the Bath for "the great contributions which, with his gifted wife, he has made to the new science of astro-physics." Lady Huggins outlived her husband by five years and died in 1915.

The Great Debate

Belief in the Stoic cosmos waned in the eighteenth, waxed once more in the nineteenth, and finally expired in the twentieth century. The so-called "Great Debate" in 1920 between Harlow Shapley and Heber Curtis, over a one-island versus a many-island universe was the last battle in the 2,000 years' war between the defenders of the rival Epicurean and Stoic systems.

At the height of the controversy over the size of the Galaxy and the distances of nearby extragalactic systems, some astronomers, notably Shapley at the Mount Wilson Observatory, favored an updated version of the one-island universe popular in the nineteenth century. This version of the universe now consisted of a supergiant Galaxy surrounded

INTERIOR OF THE OBSERVATORY, 1860—1869.

10.2 The laboratory invades the observatory. "The observatory became a meeting-place where terrestrial chemistry was brought into direct touch with celestial chemistry," wrote William and Margaret Huggins. (*Atlas of Representative Stellar Spectra,* p. 8.)

by globular clusters and immersed in infinite and empty space. But other astronomers, notably Curtis at the Lick Observatory, favored the renascent many-island universe composed of widely separated galaxies stretching away endlessly. According to this view of the universe, the extragalactic systems we see in the form of faint nebulae are giant stellar systems like our Galaxy.[10]

The Dutch astronomer Jacobus Kapeteyn, using more stars than ever before and better statistical methods, in 1922 constructed a model of the Galaxy similar to William Herschel's model, although on a much larger scale. But parallax measurements could probe only our local region of the Galaxy. Measurements of larger distances were based on uncertain assumptions and nobody, with the exception of the American astronomer Edward Barnard and the Swiss astronomer Robert Trumpler, took seriously the problem of absorption of light by interstellar matter.

Henrietta Leavitt, an astronomer at Harvard University, opened the way for a breakthrough in the measurement of astronomical distances. In 1908, while studying certain variable stars—stars that rhythmically change their brightness—in the Magellanic Clouds (two nearby irregular galaxies), she noticed that their periods depend on their luminosities: The greater the luminosity, the longer the period. Ejnar Hertzsprung, a Danish chemist who turned to astronomy, identified these variable stars as the well-known cepheids. Unfortunately, no cepheid is near enough for its distance to be determined by parallax, and Hertzsprung resorted to less direct statistical methods. Thus, the cepheids were calibrated and Shapley developed the period–luminosity relation into a yardstick for measuring the distance of more remote cepheids. The door to the new cosmography had opened. The cepheids served as landmarks in the survey of the Galaxy, and the most luminous cepheids could then be used to determine the distances of nearby galaxies.

More than a hundred globular clusters—each a compact swarm of hundreds of thousands of old stars—are scattered around in our Galaxy. The distribution of these star clusters interested Shapley, and with the 100-inch telescope on Mount Wilson he estimated their distances by their angular sizes, their brightest stars, and the periods and apparent luminosities of their cepheid variables. He found the clusters to be spherically distributed about a distant point in the constellation of Sagittarius, and he proposed that this point must be the center of the Galaxy. Jan Oort, a Dutch astronomer, showed in 1927 that the disk of the Galaxy rotates about this center, and a little later estimated that its distance is 30,000 light-years. Shapley often said that Copernicus overthrew the geocentric system, but that he himself had overthrown the heliocentric system by showing that the Sun is not at the center of the Galaxy (Figure 10.3).[11]

In 1917 Shapley commented briefly on the riddle of cosmic darkness, referring to it as the "time-honored problem of the finiteness of the stellar universe:"[12] In a footnote he wrote, "By universe in this sense

10.3   See yonder, lo, the Galaxye
Which men clepeth the Milky Wey,
For hit is whyt.
Chaucer, *The House of Fame*

A modern illustration of the Galaxy, presenting a side-view of its rotating disk, showing the central bulge and the surrounding globular clusters. The diameter of the disk is 100,000 light-years. The Sun lies a third of the way from the center toward the rim of the disk. (Adapted from J. S. Plaskett, *Popular Astronomy* 47: 255.)

we mean the system that is formed by stars of the kind that surround the sun and appear to us in all parts of the sky. Either the extent of the star-populated space is finite, or the heavens would be a blazing glory of light." He believed that the starry cosmos consisted of a vast Galaxy of stars—a one-island universe—with midget systems within and on the outskirts, and infinite but empty space beyond. How else could one solve the riddle of darkness? He clung to his belief until the late 1920s, and was, one might say, the last of the Stoic cosmologists (Figure 10.4).

The long controversy about extragalactic nebulae finally ended in 1924 when Edwin Hubble of the Mount Wilson Observatory succeeded in resolving a few bright cepheid variables in the Andromeda Nebula. By measuring their periods and apparent luminosities, and using the

10.4   Harlow Shapley (1885–1972). (Courtesy of the Harvard University Archives. Photograph by John Brooks.)

10.5 The Pleiades in Taurus, M 45. These newborn stars appear nebulous because of the surrounding gas from which they formed. (Courtesy of Kitt Peak National Observatory, Cerro Tololo Inter-American Observatory.)

period–luminosity law of Leavitt and Shapley, he determined the distance of these stars and established that the spiral nebula in Andromeda is indeed a separate system of coordinate rank with the Galaxy.[13] Revised estimates place the Andromeda Nebula at a distance of two million light-years.

Cataloging the Nebulae

In 1781 Charles Messier (1730–1817), a French astronomer and an avid hunter of comets, compiled a catalog listing a hundred or so faint nebulosities. When comets approach the inner Solar System they at first appear as faint patches of light similar to many nebulae. Messier's compilation helped comet hunters to avoid confusing these nebulae with faint comets.[14] But his catalog scratched only the surface. William Herschel, with his improved and more powerful telescopes, increased the number of recorded nebulae to twenty-five hundred; his son, John, added several thousand more, and the number has ever since continued to grow. Nowadays nebulae are cataloged according to their classification as star clusters, gas clouds, or galaxies.

From observations made with sensitive instruments in the ultraviolet, visible, infrared, and radio wavelength regions of the spectrum, we now know that the "whitish clouds" observed by Galileo are either clusters of stars (such as Praesepe, M 44, in the constellation of Cancer) or luminous clouds of gas where star formation occurs in the disk of the Galaxy (such as the Trifid Nebula, M 20, in Sagittarius and the Pleiades, M 45, in Taurus; Figure 10.5), and that many of the "patches of similar aspect" seen away from the Milky Way and unobscured by interstellar dust are either farflung globular clusters (such as M 13 in Hercules on the outskirts of the Galaxy), or distant galaxies (such as the Andromeda Nebula, M 31; Figure 10.1).

PART THREE

THE ★ ★

RIDDLE

CONTINUES

The Fractal Universe

One of the arguments advanced in favor of the spatial extinction of light was that, if there is not such extinction, the whole heavens ought to be one blaze of solar light—admitting the universe to be infinite, because it was contended that there could then be no direction in space in which the visual ray would not encounter a star (i.e., a sun). This argument is fallacious, for it is easy to imagine a constitution of a universe literally infinite which would allow of any amount of such directions of penetration as *not* to encounter a star. Granting that it consists of systems subdivided according to the law that every higher order of bodies in it should be immensely more distant from the centre than those of the next inferior order—this would happen.

John Herschel to Richard Proctor, August 20, 1869

EMANUEL SWEDENBORG, a Swedish scientist and mystic, and more a Cartesian than a Newtonian, set the ball rolling in *Principia Rerum Naturalium* (1734) by speculating on a structured universe composed of systems enfolding systems. He imagined that the universe consisted of magnetic vortices on all scales, extending from corpuscles to the Milky Way, and said, "Nature is always the same and identical with herself." Only a few years later Immanuel Kant applied this dictum to its limit.

In Kant's hands the Newtonian universe, infinite in space, acquired not only an eternity of future time but also astronomical systems on a scale never dreamt by any previous author. The moving stars, said Kant, cluster to form galaxies. Possibly the galaxies themselves, weaving orbits about one another, cluster to form even larger systems. Perhaps, pulled hither and thither by gravity, these larger groups of galaxies cluster to form vaster systems. And perhaps these vaster groups are tossed and turned in the maelstroms of even vaster systems, and so on, without end in a hierarchical universe of incomprehensible immensity. In Kant's own words, "With what astonishment we are transported when we be-

hold the infinite multitude of worlds and systems which fill the extension of the Milky Way!"

> But how is this astonishment increased, when we become aware of the
> fact that all these immense orders of star-worlds again form but one of a
> number whose termination we do not know, and which perhaps, like the
> former, is a system inconceivably vast—and yet again but one member in
> a new combination of numbers!

"There is here no end but an abyss of real immensity, in presence of
which all the capability of human conception sinks exhausted, although
it is supported by the aid of the science of number."[1]

In his *Cosmological Letters*, published in 1761, Johann Lambert, a German mathematician, also proposed that stars conglomerate in milky
way systems and speculated on the possibility of a hierarchical or multilevel universe.[2] The systems of any level, he said, have mystical dark
centers around which cluster the centers of the next lower level, and
which themselves cluster around the centers of the next higher level.
Kant's hierarchical arrangement was evolutionary and infinite in extent,
whereas Lambert's arrangement was static and finite. Lambert claimed
with some justice that he had conceived the astonishing idea of a hierarchical or multilevel universe independently of Kant.

We talk lightly about infinite space, and words such as "infinite universe" slip off the tongue without much thought. A mental picture of
worlds on worlds forming a hierarchy of structures of ever increasing
size makes us begin to appreciate the immensity of unbounded space.
Furthermore, the hierarchical picture reveals another solution to the
riddle of darkness. This new and rather subtle solution surfaced in
the nineteenth century and for a time was energetically championed in
the early years of the twentieth century.

The New Solution

John Herschel in 1848, in his review of the first volume of Alexander
von Humboldt's *Kosmos*, hinted at the possibility of hierarchical structure as a solution to the riddle of darkness.* Even if stars are infinite in
number, he said, the sky need not blaze at every point with starlight in
the manner considered by Dr. Olbers. "Nothing is easier than to imagine
modes of systematic arrangement of the stars in space . . . in consonance

---

*Entry 8 in the table of "Proposed Solutions."

with what we see around us."[3] Without doubt he had in mind a hierarchical arrangement of the kind proposed by Kant and Lambert in which stars are clustered to form larger and larger systems.

A hierarchical or multilevel arrangement of the universe seems more realistic today than in the days of Kant and Herschel, with our discovery that stars indeed cluster to form galaxies, which in turn cluster into groups of galaxies, which in turn cluster to form superclusters, which may, for all we know, cluster to form yet vaster systems.

The hierarchical or multilevel solution is rather subtle and links with the modern study of fractal structures (Figure 11.1).[4] The forest analogy helps us to understand how hierarchy solves the riddle of darkness. Imagine that we stand in a hierarchical forest. Trees cluster to form copses, which cluster to form woods, which cluster to form larger woods, and so on (Figure 11.2). By arranging each cluster to be transparent (meaning that we see between its trees), we succeed in eliminating the continuous background, and are free to look out and see the world outside the forest. Even in a forest of infinite extent, having an infinite number of trees, we can eliminate the continuous arboreal background.

A Hierarchical Forest

Our aim is to determine whether we can look out between the trees, no matter where we stand and no matter how large the forest, and catch glimpses of the external world. Let us choose a place where the maximum number of trees intervenes between us and the outside world. If we see out of the forest at this place, then we can see out at all other places. We could start by standing in a treeless space between copses. But then more trees would intervene between us and the outside world if we stood inside a copse. Or we could stand in a treeless space between woods, or between large woods, or between larger woods, and so on, and in each case progressively fewer trees would intervene between us and the outside world.

Usually copses and woods and larger clusters of trees are irregularly shaped. For simplicity we assume that clusters of all sizes are circular. We assume also that trees everywhere are alike. We start on the smallest scale by standing in any one of the smallest clusters—a copse. We ask: What is the condition ensuring that the trees in this copse do not form a continuous background? Wherever we stand in the copse we want to look out in any direction between the surrounding trees and see other

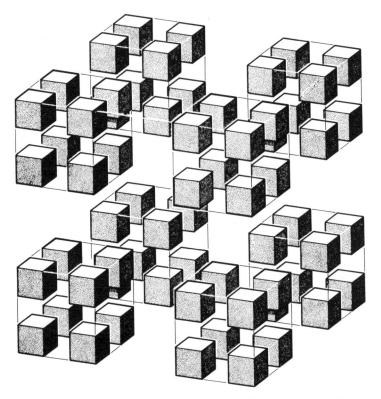

11.1   Similarity on different scales is an important feature of a fractal configuration. The irregularities of a coastline viewed on the scale of tens of meters, then kilometers, then hundreds of kilometers look often much the same. The cubic fractal in this figure exhibits similarity on different scales. In many fractals a number N is proportional to L^D, where L is the scale and D the fractal dimension. In this figure N is the number of elementary cubes, and we find that $D = 3\log2/\log 3 = 1.89$. This fractal satisfies the Seeliger–Charlier hierarchical condition for transparency.

copses. The answer is that the background limit of the copse must exceed its diameter. Consequently, the copse must contain insufficient trees to form a continuous background. The background limit, discussed earlier, equals the average area occupied by a single tree divided by the width of a tree at eye-level. Hence, when

$$\text{diameter of copse is less than } \frac{\text{area occupied by one tree}}{\text{diameter of tree}}$$

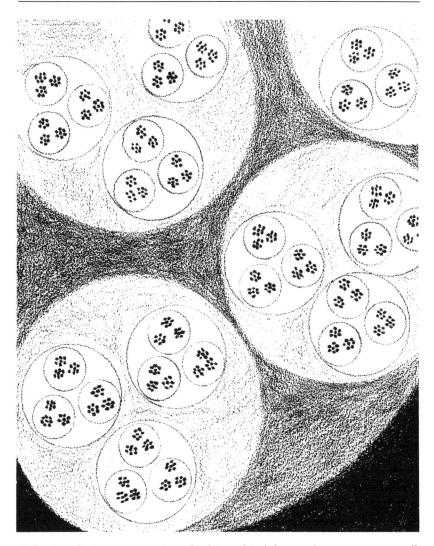

11.2   A stylistic representation of a hierarchical forest of trees, copses, small woods, large woods . . . (E. R. Harrison, *Cosmology,* courtesy of Cambridge University Press.)

we can look out between the trees of the copse from anywhere and in any direction.

When we stand in a copse in a hierarchical forest, we stand also somewhere in a wood that consists of many copses. We have established that we can look out between the trees of our copse and see other copses of the small wood. Let us suppose that all copses are similar and transparent like the one we stand in. We ask: What is the condition that the trees in this wood do not form a continuous background? Wherever we stand in the wood we want to look out in any direction between the trees and see other woods. The answer is much the same as before: the background limit of the wood must exceed its diameter. The background limit of the wood equals the average area occupied by a single tree in the wood divided by the width of a tree at eye-level. Hence, when

$$\text{diameter of wood is less than } \frac{\text{average area occupied by one tree}}{\text{diameter of tree}}$$

we can look out between the trees of the wood from anywhere and in any direction. Notice that because copses are transparent, we may for the purpose of this elementary calculation regard the trees as uniformly distributed throughout the wood.

We argue in the same way at each level of the hierarchical forest, requiring that each cluster be transparent. Any cluster, however large, fails to form a continuous background of trees when

$$\text{diameter of cluster is less than } \frac{\text{average area occupied by one tree}}{\text{diameter of tree}}$$

and in any cluster the average area occupied by one tree is the area of the cluster divided by the number of trees in the cluster. As we ascend the hierarchy to higher levels of clustering, the average area occupied by one tree gets bigger and therefore the average density of trees gets progressively less.

The area of a circular cluster is $\pi (\text{diameter})^2/4$, and our condition for seeing out of a cluster becomes:[5]

$$\text{diameter of cluster must be greater than}$$
$$(\text{number of trees in cluster}) \times (\text{diameter of tree})$$

If each tree has a diameter at eye-level of 1 meter, and a wood contains 1,000 trees, we look out between the trees and see the forest beyond when the diameter of the wood exceeds 1 kilometer. This result is independent of how the wood subdivides into transparent copses.

Our conclusion that the size of a cluster must exceed the width of a tree multiplied by the number of trees in the cluster is quite general and applies to copses, woods, large woods, and so on, even to the forest itself. When this condition is satisfied on all scales of clustering, the forest at every point fails to present a continuous background of trees.[6]

A Hierarchical Universe

Hitherto, our main problem has been that every line of sight in an endless star-filled universe eventually encounters the surface of a star. Consequently stars cover the sky. The forest analogy shows that a hierarchical arrangement invalidates the line-of-sight argument, as John Herschel anticipated.

For simplicity we assume that clusters on all scales are spherical in shape, and stars everywhere are alike. Let us start at the level in which stars are clustered into galaxies. An observer looks out between the stars in any direction from anywhere and sees external galaxies when the diameter of the galaxy is less than its background limit. This means

$$\text{diameter of galaxy is less than } \frac{\text{volume occupied by one star}}{\text{area of star}}$$

We showed in an earlier discussion that the absorption of starlight generally fails to solve the riddle of darkness, and therefore in the present treatment we may neglect interstellar absorption by gas and dust. The "volume occupied by one star" is the volume of the galaxy divided by the number of stars in the galaxy. A spherical galaxy has a volume $(\pi/6)(\text{diameter of galaxy})^3$, and the above condition that we see between the stars of the galaxy becomes:[7]

$$\text{(diameter of galaxy)}^2 \text{ must be greater than}$$
$$\text{(number of stars in galaxy)} \times \text{(diameter of a star)}^2$$

Thus, if a galaxy contains a trillion stars and the typical diameter of a star is 5 light-seconds, then an observer looks out between the stars when the diameter of the galaxy exceeds 5 million light-seconds, or 2 light-months. Most galaxies have diameters of tens and hundreds of thousands of light-years and are therefore transparent when irrelevant interstellar absorption is ignored.

As in the case of a forest, we can generalize and say that a cluster at any hierarchical level is transparent when:

$$\text{(diameter of cluster)}^2 \text{ must be greater than}$$
$$\text{(number of stars in cluster)} \times \text{(diameter of a star)}^2$$

We construct, step by step, a hierarchical universe in which stars gather into transparent (meaning optically thin) galaxies, galaxies into transparent clusters, clusters into transparent superclusters, and so on. By proceeding in this way on larger and larger scales, we construct an endless hierarchical universe in which stars fail to cover the sky and the sky at night remains dark.[8]

A hierarchical universe has the intriguing property that its density gets less when averaged over progressively larger volumes and ultimately approaches zero in a universe of infinite extent having an infinity of levels.

Carl Charlier

John Herschel proposed the hierarchical solution to the riddle of darkness; and Richard Proctor, an English astronomer and a popularizer of science, who emigrated to America the year he married at age forty-four in 1881, presented in *Other Worlds than Ours* (1871) a semiquantitative treatment of a hierarchical solution. But neither Herschel nor Proctor derived the conditions with any generality and precision.

Then Fournier d'Albe, whom we have already met and shall meet again later, in his *Two New Worlds* championed the notion of hierarchy. Among his perceptive comments we find, "William Herschel remarked that he had discovered fifteen hundred universes. He referred, of course, to as many nebulae, which he believed to be galaxies external to our own. But . . . it is risky to use the word universe in the plural number without some special definition. If by universe we mean the totality of things, there can, of course, be only one totality."[9] He elaborated on the idea that our visible universe is just one in a multiuniverse. All universes have similar structure, he suggested, and differ only in scale. In the universe immediately below, the infraworld, everything is 10 billion trillion times smaller, and in the universe immediately above, the supraworld everything is 10 billion trillion times larger. Our atoms are the solar systems in the infraworld and our solar systems the atoms in the supraworld (Figure 11.3).

At first, Carl Charlier, a professor at the Swedish University of Lund and director of the University Observatory, subscribed to the accepted cosmological model of our Galaxy immersed in infinite and empty space, for it offered the only reasonable explanation of darkness at night. After reading Fournier d'Albe's book, he changed his mind and began to champion the hierarchical model.

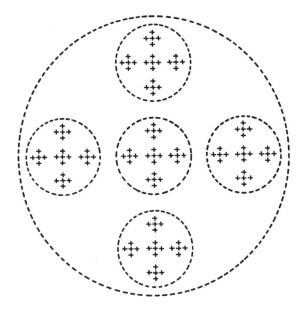

11.3 Adapted from Fournier d'Albe's diagram of a hierarchical universe, or multiuniverse, in his *Two New Worlds* (1907). This diagram, he wrote, shows "that an infinite series of similar successive universes may exist without producing a 'blazing sky.'" When the number of stars in each cluster increases with the radius of the cluster the "sky will always appear quite black."

First in 1908, then in 1922 in a more detailed paper on "How an infinite world may be built up," Charlier derived a hierarchical solution to the riddle of darkness at night similar to that described in the previous section.[10] He also showed that the Bentley–Newton gravity riddle in an infinite universe of uniformly distributed matter (known formally as the Dirichilet problem) could be solved in a hierarchical universe. A few astronomers in the early years of the twentieth century, attracted by the idea of a many-galaxy universe, regarded hierarchy as an appealing solution to the cosmic riddles of gravity and darkness.

Doubts Concerning Hierarchy

Svante Arrhenius, also a Swedish scientist, a Nobel prize-winner, and director of the Nobel Institute of Physical Chemistry, in an essay on "The infinity of the universe," wrote in 1911, "In modern times it is assumed by many astronomers that the universe is finite and surrounded by an infinite empty space into which the sun and stars radiate an energy forever to remain lost. Frequently also the idea is voiced that our sun occupies a position near the center of such a finite universe."[11] Arrhenius argued against this anthropocentric one-galaxy picture of the universe and also against the hierarchical picture. The evolving universe, he said, though infinite in space, is finite in age, and the star-covered sky is dark because of obscuration caused by nonluminous bodies such as cosmic dust, meteorites, planets, and the unseen companions of stars. The absorption solution advocated by this famous scientist had in fact already been disproved.

Astronomical hierarchy succeeds in making the sky dark by pushing the background limit out beyond any size we wish to assign to the universe. If, as is usually assumed by the proponents of the hierarchy solution, the universe is infinite in size and contains an infinity of stars, the sky stays dark at night because the background limit has been pushed out to infinite distance. But a background limit pushed out to infinite distance requires a hierarchy consisting of an infinite number of levels. A *finite* number of levels—stars, star clusters, galaxies, galaxy clusters, superclusters—fails to solve the riddle, because a uniform distribution of superclusters extending to infinity will still create a bright starlit sky. The hierarchy solution to the riddle of darkness has therefore a serious drawback: It requires that clustering occur on progressively larger scales, escalating to infinity in an endless universe.

Hierarchy has a further drawback. Like the other solutions we have encountered, it ignores the speed of light and assumes that all parts of the universe, no matter how remote, are open to inspection.

TWELVE

The Visible Universe

But the probability amounts almost to certainty that star-strewn
space is of measurable dimensions. For from innumerable stars
a limitless sum-total of radiations should be derived, by which
darkness would be banished from our skies; and the "intense
innane," glowing with the mingled beams of suns individually
indistinguishable, would bewilder our feeble senses with its mo-
notonous splendour. This laying bare, so to speak, of the empyrean would be the simple and certain result of the continuance
of sidereal objects comparable with that prevailing in our
neighborhood.

Agnes Clerke, *The System of the Stars*

THE IDEA that we see things in the external world instantaneously dates
back to prehistory and in the ancient world was associated with the
visual ray theory.[1] According to this theory, as explained by the textbook
writers Euclid and Ptolemy at the Museum in Alexandria, the eye emits
visual rays—Platonic "divine fire"—and these probing rays, reflected
by things in the external world, return to the eye with visual information. To this day we think and speak automatically in terms of this radar
theory, in which the eye acts as an emitter of visual rays: "She cast a
piercing look," or "He returned a smiling glance."

To the ancients and medievalists, the demonstration that visual rays
travel at infinite speed could hardly have been more convincing. Turn
to the heavens at night with closed eyes, said the Platonists, now open
them, and instantly you will see through treetops the wispy clouds, the
shining Moon, and the twinkling stars, all at greatly different distances.
Does this not prove beyond the least doubt that visual rays and indeed
all light rays travel at infinite speed?

Galileo in his *Two New Sciences* in 1638 described an experiment to
determine whether light travels at finite speed. Let two experimenters

with covered lanterns stand at some distance from each other, he said. One experimenter uncovers his lantern, and the other, on seeing this sudden light, instantly uncovers his lantern. A delay observed by the first experimenter would indicate that light travels at finite speed. "In fact I have tried this experiment only at short instances, less than a mile, from which I have not been able to ascertain whether the appearance of the opposite light was instantaneous or not; but if not instantaneous it is extraordinarily rapid."[2]

Reason assured Descartes that light travels at infinite speed. By regarding the Earth and Moon as two lanterns, he argued that when the Earth eclipses the Moon we observe the Earth's shadow on the Moon in a direction precisely opposite to that of the Sun. This could not occur, he pointed out, if light took as little as even one hour to travel from the Earth to the Moon.[3] Consider how confused our reconstruction of the external world would be, said Descartes, if the rays composing images in the eye and coming from different distances were emitted at different times. Indeed, if light propagated at less than infinite speed, then he would admit that the whole of his philosophy had been shaken to its foundations. The thought that when we look out in space we also look back in time seemed to Descartes, and to many Cartesians who followed, too incredible to be taken seriously. Neither the ancients, nor Galileo, nor Descartes realized how great, but not infinite, is the actual speed of light.

Light, wrote Robert Hooke in the *Micrographia*, propagates in all directions with equal speed, and "every *pulse* or *vibration* of the luminous body will generate a Sphere, which will continually increase, and grow bigger, just after the same manner (though indefinitely swifter) as the waves or rings on the surface of the water swell into bigger and bigger circles about a point of it, where, by the sinking of a Stone the motion was begun." Hooke discussed the difficulty of measuring the very small interval of time taken by light traveling from one place to another. Even if "we should grant the progress of light from the Earth to the Moon, and from the Moon back to the Earth again to be full two Minutes in performing, I know not any possible means to discover it."[4] His two minutes, though much less than Descartes's two hours, still greatly underestimated the actual speed of light. The time taken by light to reach the Moon is not one minute but slightly more than one second. But this inventive man, who regarded all heat as atomic motion and light as transverse vibrations, by 1680, in "Lectures of light," had changed his mind and come to the conclusion that the propagation of light through the "immense *Expansum*, as far we can yet find, is made in a Point or Instant of time; and at the very Instant that the remotest Star does emit Light, in that very Instant does the Eye upon the Earth receive it, though it be many Millions of Millions of Miles distant, so that in Probability no time is spent between the emission and Reception."[5]

Io, Princess of Argos

Gian Cassini at the University of Bologna was renowned for his accomplishments as an astronomer. In 1669 he accepted an invitation from the Sun King (Louis XIV) to join the staff of the new Paris Observatory, where he soon became virtual director. He obtained French citizenship, married an heiress, and founded the Cassini dynasty of astronomers. He continued his studies of the revolutions and rotations of the planets, and compiled records of the eclipses of Galileo's four moons of Jupiter. The innermost of these moons, Io, had an orbital period around Jupiter sufficiently short (1 day, 18 hours, 28 minutes) to serve as a celestial clock of interest to mariners in their attempts to determine longitude at sea.

Cassini noticed that for a few months, while the Earth approached Jupiter, the eclipses of Io by Jupiter occurred a few minutes early, and for a few months, while the Earth receded from Jupiter, the eclipses occurred a few minutes late. He considered the possibility that these variations in Io's period are caused by a finite speed of light. Light reaches us sooner when the Earth is nearer to Jupiter, and later when farther away, thus causing the observed variation in the eclipses. But being a Cartesian, and sharing Descartes's fears of a finite speed of light, he hastily withdrew his suggestion.

The French astronomer Jean Picard, after an expedition to determine the longitude of Tycho Brahe's old observatory on the Island of Hveen, returned to Paris in 1672 accompanied by the young Danish astronomer Ole Roemer. Roemer was appointed tutor to the Dauphin and assigned the task of assisting in updating the occultation records of Jupiter's moons. Like Cassini, Roemer noticed the irregularities in Io's eclipses. The satellite's immersions (disappearances) behind Jupiter and emersions (reappearances) rarely occurred precisely on time. With improved data, he became convinced that finiteness of the speed of light explained these variations. In September 1676 he announced at a meeting of the Paris Academy of Sciences that the eclipse of Io on November 9 would occur 10 minutes late. He went around repeating this prediction, explaining that the Earth was moving away from Jupiter, and the delay would be caused by the extra distance light had to travel to catch up with the Earth. His successful prediction, followed by his paper read before the Academy, established him as the discoverer of the finite speed of light. He pointed out that the change that had occurred in Io's orbital period, though "not sensible in two revolutions, became very considerable in many taken together."[6]

Five years later, on returning to Copenhagen, Roemer was appointed a professor at the university; he became mayor of Copenhagen, and then astronomer royal of Denmark. Among his many inventions must be included the mercury thermometer in the form used nowadays. Gabriel Fahrenheit, a German-Dutch instrument maker, only copied Roemer's method of construction and calibration (ice at 32 degrees and boiling water at sea-level at 212 degrees); the misnamed Fahrenheit thermometer should be called the Roemer thermometer.

The Cosmic Consequences

Most Newtonians, with the notable exception of Hooke, accepted Roemer's result. Halley reviewed the updated data on Io's eclipses in 1694 and came to the conclusion that light travels an astronomical unit (the Sun–Earth distance) in a time of 8 minutes 30 seconds, a result only 10 seconds more than the present-day value. Also in 1694 Francis Roberts, a musician interested in the theory of sound and a fellow of the Royal Society, published a paper, "Concerning the distance of the fixed stars," in which he used the speed of light to indicate how vast are the distances to the nearest stars. From the fact that "sensible Parallax to the fixt Stars" had not been detected, he reckoned that "Light takes up more time in Travelling from the Stars to us, than we in making a *West-India* Voyage (which is ordinarily performed in six Weeks)."[7]

Cartesian opposition to the finite speed of light crumbled in 1729 when James Bradley's discovery of the aberration of light vindicated Ole Roemer.[8] Bradley discovered that stars move backward and forward through very small angles during the year because of the Earth's orbital motion around the Sun. When walking in vertically falling rain, we notice how the rain slants toward the face, and for this reason an umbrella is tilted forward. Light slants in a similar way to a moving observer, and this effect is referred to as aberration (an odd name introduced in 1737 by the French mathematician Alexis Claurant).

Little further serious interest was taken in the properties of light until the beginning of the nineteenth century, when Thomas Young[9] developed the wave theory to explain interference and diffraction. Then, in the following decades, advances were made in rapid succession; Armand Fizeau and Jean Foucault determined more exactly by terrestrial methods the speed of light, and Michael Faraday, James Clerk Maxwell, and other physicists founded the electromagnetic theory of light that unified electricity and magnetism (Figure 12.1). An electrical industry proliferated, and in the first half of the twentieth century radio communication linked all parts of the globe. The technologies of an electronic age had emerged.

At first the cosmic consequences of a finite speed of light received little attention. Many astronomers remarked that light can travel only a finite distance in a finite period of time.[10] A few tried circumspectly to explain in popular works that in a universe of infinite space and finite age we observe only a finite portion of space; but their words, guarded and interlarded with expressions of piety, were too obscure to make impact. Most astronomers, however, went little further than to express wonderment that we observe astronomical bodies as they were thousands and even millions of years ago. The implication—that the universe must be at least as old as the time taken by light to travel from the farthest visible bodies—received little comment in books and lectures, possibly because it ran counter to popular religious opinion.

Descartes's fear of the confusion caused by a finite speed of light lingered on. A finite speed means that an object at the moment of observation is not what it seems to be; it has not only moved elsewhere but has also changed its appearance. A finite speed means that the observable portion of the universe—which I shall refer to as the *visible universe*—extends no farther than the distance that light has traveled since the beginning of the universe. The implication of this revolutionary concept—that the age of the universe cannot be less than the extent of the visible universe divided by the speed of light—took quite a while to sink in. The alarming thought that when we look out in space to the limit of the visible universe we see things as they were in the beginning was rarely expressed or discussed in pre-relativity days (Figure 12.2).

Two Million Light-Years

William Herschel occasionally remarked that we see the heavens as they were long ago. In 1802 he reported that the rays of light from a certain nebula must have been "almost two millions of years on their way; and

12.1   Light brighter than any made before by human beings. Humphry Davy demonstrated the electric arc at the Royal Institution in 1808. For his work on electricity Napoleon awarded Davy a medal in 1806, despite the war between France and England. Davy accepted the medal on the grounds that science should be above national divisions. Michael Faraday, hired as a youthful "bottle-washer," succeeded Davy in 1825 and became the foremost experimenter of his time. When William Gladstone, Chancellor of the Exchequer, on a visit to the Royal Institution, asked Faraday what good was electricity, Faraday replied, "One day, Sir, you may tax it." (By A. R. Thomson, in F. Sherwood-Taylor, *An Illustrated History of Science,* London: Heinemann.)

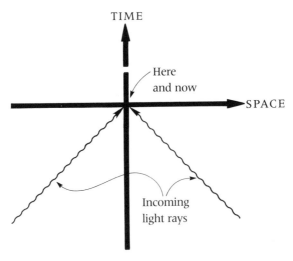

12.2 On looking out in space the observer looks back in time. This space and time diagram shows rays of light arriving at the observer (distance is measured in light-travel time).

that, consequently, so many years ago, this object must already have had an existence in the sidereal heavens, in order to send out those rays by which we now perceive it."[11] John Herschel, in *A Treatise of Astronomy* (1830), followed this theme cautiously: "Among the countless multitude of such stars visible in telescopes there must be many whose light has taken at least a thousand years to reach us; and that when we observe their places and note their changes we are, in fact, reading only their history of a thousand years' anterior date, thus wonderfully recorded."[12]

Such remarks by the Herschels inspired many authors to comment about distant worlds seen as they were long ago. Contrary to our misconceptions of the nineteenth century, the boldest comments came not from the laity but from astronomers who had taken orders. John Nichol, schoolmaster, minister, and then regius professor of astronomy at the University of Glasgow, who introduced the word "evolution" into astronomy, rejoiced in the immensity of the universe in his "series of letters to a lady" in 1838, entitled *The Architecture of the Heavens*.[13] In a revised edition, he wrote in 1850 that light travels across abysses whose immensity stuns the imagination. Nichol ranked among the diminishing number who believed that the faint nebulae were distant island universes. Their dimness informs us, he said, that they emitted the "light we are now receiving at an epoch farther back into the past than this momentary epoch of our human race by about thirty millions of years!" Another Scottish teacher and minister, Thomas Dick, author of scientific and philosophical works of considerable popularity in Britain and America,[14] wrote in *The Sidereal Heavens* (1840), "Such distances are amazing and almost terrifying to the human imagination." He dared to think that "A seraph might wing his flight with the swiftness of light for millions of years through the regions of immensity, and never arrive at a boundary."

Alexander von Humboldt, a German scientist of broad intersts who died at ninety in 1859, referred to William Herschel's remark of 1802 in his book *Cosmos:*[15]

> The aspect of the starry heavens presents us with the spectacle of that which is only apparently simultaneous, and however much we may endeavour by the aid of optical instruments to bring the mildly-radiant vapor of nebulous masses or the faintly-glimmering starry clusters nearer, and diminish the thousands of years interposed between us and them, that serve as a criterion of their distance, it still remains more than probable, from the knowledge we possess of the velocity of transmission of luminous rays, that the light of remote heavenly bodies presents us with most ancient perceptible evidence of the existence of matter.

Humboldt discussed the dark night-sky riddle but failed, despite these comments, to infer from the finite speed of light a limit to the extent of the visible universe of stars.

Richard Proctor explained in 1874, in an essay entitled "The flight of light," that with telescopes the "range of time over which our vision extends is enormously increased, and it is certainly not too much to say that some of the fainter stars revealed by the great Rosse telescope lie at distances so enormous that their light has taken more than a hundred thousand years in reaching us." He continued, "Every moment light reaches this earth from unseen orbs so far away that the journey over the vast abysses separating us from them has not been completed in less than millions of years."[16]

Robert Ball, royal astronomer of Ireland, in his 1892 book, *In Starry Realms,* said of certain distant stars, "The light they emit has been on its journey for 18,000 years before it reached us. When we look at those lights tonight we are actually viewing them as they were 18,000 years ago. In fact, those stars might have totally vanished 17,000 years ago, though we and our descendents may still see them glittering for yet another thousand years."[17]

These quotations from popular astronomical and scientific works in the nineteenth century indicate how widespread was the realization that when we look out in space we also look back in time. But only one author, a poet, associated this fact with the riddle of cosmic darkness.

The Golden Walls of Edgar Allan Poe

How distant some of the nocturnal suns!
So distant, says the sage, 'twere not absurd
To doubt that beams set out at Nature's birth
Had yet arrived at this so foreign world,
Though nothing half so rapid as their flight!

Unknown author, quoted in John Nichol,
Views of the Architecture of the Heavens

THE EARLY Victorians were moderately tolerant of their amateur scientists. Indeed, many investigators in the natural sciences, of independent means, whom we would nowadays regard as amateurs, became esteemed members of learned societies. The gathering complexity of natural science and its specializations had yet to erect insurmountable barriers against nonprofessional contributors. The first clear and correct solution to the riddle of darkness, though only qualitatively expressed, came from Edgar Allan Poe, the renowned poet, essayist, critic, and amateur scientist (Figure 13.1).

In the June 1845 issue of the *United States Magazine and Democratic Review,* Edgar Allan Poe published a moving essay, "The Power of Words," in which he wrote,

> Look down into the abysmal distances!—attempt to force the gaze down the multitudinous vistas of the stars, as we sweep slowly through them thus—and thus—and thus! Even the spiritual vision, is it not at all points arrested by the continuous golden walls of the universe?—the walls of the myriads of the shining bodies that mere number has appeared to blend into unity?[1]

In February 1848, three years after referring to the golden walls of the universe, and only a year before he died at age forty, Edgar Allan

13.1   Edgar Allan Poe (1809–1849).

Poe delivered a two-hour lecture "On the cosmogony of the universe" before a scant audience at the Society Library, New York, while the Rev. Dr. John Pringle Nichol, regius professor of astronomy at the University of Glasgow, also happened to be lecturing in New York. Nichol's *Views of the Architecture of the Heavens,* bringing to the attention of the public the astounding discoveries of contemporary astronomy, and written in a style of religious humility incapable of giving offense, had caused an immense stir in the English-speaking world. Undoubtedly, Edgar Allan Poe was greatly influenced by Nichol's popular work.[2] Later in that year Poe amplified his lecture and published it as an essay entitled *Eureka: A Prose Poem*.[3] In this imaginative masterpiece, dedicated to Alexander von Humboldt, he formulated his most daring cosmological ideas. He visualized a universe rhythmically expanding and collapsing with each pulse of the Heart Divine. In an apocalyptic vision he foresaw the collapse of the present cosmic era: "Then, indeed, amid unfathomable abysses, will be glaring unimaginable suns." Of this work he wrote in a letter, "What I have propounded will (in good time) revolutionize the world of Physical and Metaphysical Science."[4]

Eureka failed to revolutionize the world of physics and metaphysics; its science was too metaphysical and its metaphysics too scientific for contemporary tastes. It constitutes, however, a most interesting and important contribution to cosmology, and only twenty-five years after Olbers wrote his paper on the riddle of darkness, it contains the first anticipation of a formally correct solution. In *Eureka* Poe wrote,

> Were the succession of stars endless, then the background of the sky would present us an uniform luminosity, like that displayed by the Galaxy—*since there could be absolutely no point, in all that background, at which would not exist a star.* The only mode, therefore, in which, under such a state of affairs, we could comprehend the *voids* which our telescopes find in innumerable directions, would be by supposing the distance of the invisible background so immense that no ray from it has yet been able to reach us at all.

The speed of light and the age of stars had at last come together to reveal a new aspect of an old problem.

In the twentieth century we have grown accustomed to the idea that our vision slices through space and time. When we gaze at the night sky, looking far out in space, we are fully aware that we see the apparitions of long ago. We find it difficult to understand why Descartes and other philosophers once viewed with alarm the prospect of splicing space and time together. Yet even to us, accustomed to the idea of looking out in space and back in time, the thought that at the horizon of the visible universe lies the creation, unveiled and open to inspection, comes as a shock.

Our vision extends a limited distance in a universe in which stars have been shining for a limited period of time (Figure 13.2). The starry

Edge of observable
universe

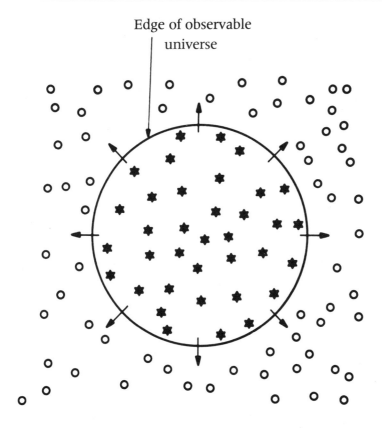

✷   Stars visible

•   Stars not yet visible

13.2   In a static Newtonian universe, uniformly populated with stars, we look out and see stars extending to a distance of roughly 10 billion light-years. At greater distances we look back to a time before the stars became luminous. The outer boundary of the sphere of visible stars—the horizon or edge of the visible universe—recedes at the speed of light. (E. R. Harrison, *Cosmology,* courtesy of Cambridge University Press.)

universe may be infinite in space, or if not infinite then vast beyond measurement, and yet the part of it that we can see—the visible universe—is comparatively small, generally much too small to contain enough stars to cover the sky.

Poe hesitated: "That this *may* be so, who shall venture to deny?" He granted that the astronomical evidence favored a one-island universe of a single system of stars, but felt attracted to a many-island universe of multiple systems. May we not infer from the evidence, he said, that this

> perceptible Universe—that this cluster of clusters—is but one of a *series* of clusters of clusters, the rest of which are invisible through distance—through the diffusion of their light being so excessive, ere it reaches us, as not to produce upon our retinas a light-impression—or from there being no such emanation as light at all, in those unspeakably distant worlds—or, lastly, from the mere interval being so vast, that the electric tidings of their presence in Space, have not yet—through the lapsing myriads of years—been enabled to traverse that interval?

A few lines later: "Let me declare, only, that as an individual I myself feel impelled to the *fancy*—without daring to call it more—that there *does* exist a *limitless* succession of Universes, more or less similar to that of which we have cognizance." By universes I assume he meant galaxies. A few pages later, he added, "but the considerations through which, in this Essay, we have proceeded step by step, enable us clearly and immediately to perceive that *Space and Duration are one.*"

From the works of William and John Herschel, John Nichol, and Thomas Dick, and the recent reviews of the first volume of Alexander von Humboldt's *Kosmos,* which described the architecture of the universe and drew attention to the long periods of time taken by light to traverse interstellar space, Poe came to the conclusion that a possible solution to the riddle of darkness is that the light of the "golden walls" has yet to reach us.* When we look far out in space we look far back to a time before the birth of stars.

The Receding Horizon

Edward Fournier d'Albe wrote a series of seven essays on "The infra-world" from September to October in 1906, and a second series of six essays on "The supra-world" from March to May in 1907 in *The English Mechanic and World of Science and Art.* These essays were collected and

---

*Entry 9 in the table "Proposed Solutions."

published with slight revisions in his book *Two New Worlds* in 1907.

The *English Mechanic* catered to all interests: metal working, carpentry, model building, how to make your own electric motor, how to build your own turbine, how to sail a yacht, how in fact everything works, from sewing machines to blast furnaces, from rainbows to stars, and all the rest, including the universe. No English person, man or woman, could afford not to subscribe. By chance, on page 202 of the April 6, 1907, issue I saw, a year or so ago, a footnote referring to a paper by Lord Kelvin in the *Philosophical Magazine*. On tracing that long-forgotten paper by Kelvin, I stumbled on a discovery that forms the basis of the next chapter. As I continued to turn the musty pages of the *English Mechanic*, noticing the surprising depth of its catholic science and the unhesitating use here and there of mathematics, nostalgia overcame me; I was precipitated back into an age whose last stages I can only vaguely recall, an age that was quite unlike the present with its glossy and comparatively superficial science magazines.

In the first essay of "The supra-world," in the March 29 issue of 1907, Fournier d'Albe tackled the problem of the "blazing-sky theory." He noted the work of various authors, including Simon Newcomb and Agnes Clerk, and then remarked, "There is, of course, another possibility. If the world was created 100,000 years ago, then no light from bodies more than 100,000 light-years away could possibly have reached us up to the present." This idea came not from Edgar Allan Poe, but presumably from Lord Kelvin, whose work he referred to briefly in the next essay published the following week. He then commented on an aspect of the visible universe not previously discussed (I quote from his book): "But light from stars farther and farther away would be continually arriving at the earth's surface, and our vision into space, confined at present by the Milky Way, would be expanding at the rate of 186,000 miles per second."[5]

If the universe began in the finite past, then only a finite part is visible; the rest—the part beyond a certain distance—cannot be seen because the light from this part has not had time to reach us. But as the universe ages we see more of it. The visible universe is bounded by a horizon that recedes at the speed of light, and the part accessible to observation grows ever larger. This expansion of the horizon of the visible universe must not, of course, be confused with the dynamical expansion of the whole universe discovered a few years later.

Each observer possesses a personal visible universe consisting of the surrounding region of visible stars. An observer exists at the center of his or her own visible universe. When observers move around, traveling figuratively from place to place in the universe, they take with them their own visible universes. This is like the familiar visible region of the Earth's surface, which is bounded by a horizon that moves around with each observer. A visible universe stretches away from an observer to a horizon at a distance determined by the speed of light multiplied by the time elapsed since the firstborn stars began to shine. Thus, if stars began to shine a zillion years ago, the horizon lies at a distance of a zillion light-years. We may suppose on the one hand that the star-filled universe began a zillion years ago, or on the other hand that stars began to shine a zillion years ago in a universe previously starless. In either case, the light from the horizon reveals the condition of the universe at the time when the first generation of stars began to shine. This argument applies in a static universe; in an expanding universe the situation is similar but slightly more complicated.

Probably most galaxies were formed a billion or so years after the big bang. The oldest stars now shining in our Galaxy were born long ago shortly after the formation of the Galaxy. Probably therefore the oldest stars in most galaxies have much the same age. If we suppose that the galaxies originated 10 billion years ago, which is roughly the age of the Milky Way, we shall not be wildly wrong. In that case, the oldest stars everywhere in the universe have also much the same age (Figure 13.3).

Homogeneity with a Vengeance

As we saw in Chapter 11, Fournier d'Albe was mainly interested in formulating a hierarchical theory of his own. "The root-hypothesis of these articles, is, however, that this world of ours is a good average sample of the universe, as it always has existed and always will continue to exist, and that, however high we ascend, or however low we descend in the scale of magnitude, we may hope to find conditions not alarmingly different from those which we have here and now learnt to know and to adapt to."[6] Here is homogeneity with a vengeance—outreaching the fondest hopes and boldest designs of all other cosmologists from ancient times to the present day. The infinite universe presents, he suggested, the same appearance not only everywhere in space, not only at every instant in time, but also on all scales from the infinitely small to the infinitely large. In a universe eternally the same, the stars shine forever, and hence the horizon of the visible universe lies at infinite distance. The blazing-sky theory therefore posed a real problem that only hierarchy seemed able to solve.

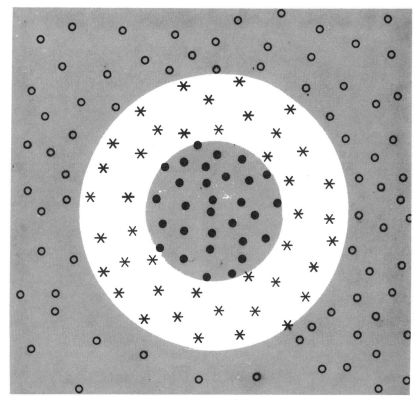

● Dead stars
� Stars visible
○ Stars not yet visible

13.3 If we wait long enough, the stars around us die out. We shall then be surrounded by an expanding dark sphere of dead stars; beyond the dark sphere will lie an expanding shell of luminous stars; and beyond this luminous shell will extend a dark universe of unborn stars.

Fournier d'Albe's fractal theory of the universe, described in *Two New Worlds*, greatly influenced the Swedish astronomer Charles Charlier, as we have seen, who became an enthusiastic advocate of unlimited astronomical hierarchy. Charlier derived the mathematical conditions for a hierarchical solution to the riddle of darkness. It is not obvious why Charlier, an able theoretician and an international figure in astronomy, took this course. Fournier d'Albe had fully and fairly shown that the visible universe is of finite size in a universe of finite age, and had shown by implication that, as a consequence, too few stars cover the sky, thus cutting the ground from under the hierarchical solution of the riddle. Why, then, go to the bother of clustering into a hierarchical system an infinity of stars that lie outside the horizon of the visible universe when they are invisible and cannot create a bright starlit sky? Either Charlier failed to grasp the implications of a finite visible universe, or he shared Fournier d'Albe's belief in a steady-state universe forever renewing itself and eternally the same.

Lord Kelvin Sees the Light

And so the lectures went on from day to day with delightful discursiveness. Here and there a reference to Stokes—"I always consult my great authority, Stokes, whenever I get a chance"— or a hit at the "brain-wasting perversity of insular inertia which still condemns British engineers to reckonings of miles and yards and feet and inches, and grains and pounds, and ounces and acres"—or an ejaculation that "there are no paradoxes in science!" Questions by the hearers set the lecturer off on new trains of thought. Discussions started at the lecture were continued at the supper table. The whole seemed one animated conference. But the continuous discursiveness of the lecturer became ominous. "How long will these lectures continue?" asked President Gilman one day of Lord Rayleigh, while walking away from the lecture-theatre. "I don't know," was the reply; "I suppose they will end some time, but I confess I see no reason why they should."

Silvanus Thompson, *The Life of William Thomson, Baron Kelvin of Largs*

WILLIAM THOMSON (Figure 14.1), later Lord Kelvin, one of John Nichol's university students, became professor of natural philosophy at Glasgow University at age twenty-two in 1846 and occupied this chair until his retirement at age seventy-five, whereupon he then, characteristically, enrolled as a research student.[1] From 1848, when he introduced the absolute (Kelvin) scale of temperature, this remarkable Victorian scientist made numerous important advances in theoretical and experimental physics. The queen knighted him in 1866 for his scientific work in connection with the first Atlantic telegraph cable and made him a peer of the realm in 1892.

Lord Kelvin (then Sir William Thomson), by invitation of President

14.1   William Thomson, Lord Kelvin (1824–1907).

Gilman on behalf of the Trustees of Johns Hopkins University, in 1884 delivered a course of lectures to a select audience composed mainly of American physicists. Albert Michelson and Edward Morley (who collaborated together later in a famous experiment to try and determine the Earth's motion through the ether) were among the audience, with Lord Rayleigh and George Forbes attending from England. The lectures—bearing the title "Molecular dynamics and the wave theory of light"—were presented extempore by Kelvin and stenographically recorded by Arthur Hathaway.[2] The greatly revised and extended lectures, with twelve appendixes, were finally published by Kelvin in 1904 in a volume bearing the title *Baltimore Lectures*.[3]

Lecture 16 in the new edition is particularly interesting. Of its twenty-two sections, the first seven follow the 1884 treatment and recall Kelvin's earlier work on the nature and density of the ether. (Until the twentieth century the luminiferous ether was thought essential for the propagation of light.) Sections eight to ten, dated November 1899, dwell on the gravitational properties of the ether. The remaining undated twelve sections, presumably written subsequently but before August 1901, deal with astronomical topics such as the size and mass of the Galaxy and the equilibrium velocities of the stars. Sections eighteen and nineteen (reproduced as Appendix 5 of this book) examine the problem of a dark starlit night sky and show that the observed state of darkness was consistent with the size of the Galaxy then in vogue and the number of contained stars. This revised lecture was published separately in August 1901 in the *Philosophical Magazine* as a paper entitled, "On ether and gravitational matter through infinite space." It was a reference to this paper in the April 6, 1907, issue of the *English Mechanic* that recently drew my attention to Kelvin's contribution to the riddle of darkness. The absence of this paper in Kelvin's *Mathematical and Physical Papers* and in the bibliography of his published work in Silvanus Thompson's *Life of William Thomson* helps to explain its total neglect in all subsequent discussions of "Olbers's paradox."[4]

Kelvin's Analysis

Kelvin saw to the heart of the problem. With some numbers and a few calculations, he solved the riddle of darkness at night quantitatively and correctly in the framework of a transparent, uniform, and static universe.

His calculations, in simplified form, show that

$$\text{the fraction of sky covered by stars} = \frac{\text{size of visible universe}}{\text{background limit}}$$

The "size of the visible universe" in this equation was at first taken to be the radius of our sidereal system, or Galaxy; the "background limit"—the average distance of visible stars in a universe uniformly populated with stars—was calculated roughly in the manner shown in Chapter 7. Kelvin's calculations ignored the possibility that the size of the visible star-filled universe might be larger than the background limit. A more general calculation, however, gives essentially the same answer, and shows that the fraction of the sky covered by luminous stars approaches unity as the size of the visible universe approaches and then exceeds the background limit.[5]

Kelvin referred to Simon Newcomb's *Popular Astronomy* illustrating William Herschel's view of the Galaxy, according to which we occupy the center of a sidereal system consisting of a billion sunlike stars and having a radius of 1,000 parsecs (Figure 14.2).[6] (One parsec equals 3.26 light-years, and is the distance of a star having a parallax of one arc second.) For convenience Kelvin also assumed that all stars are similar to the Sun and are distributed uniformly with separations much the same as in our part of the Galaxy. He then showed that the fraction of the sky covered by stellar disks was less than one-trillionth, and commented, "This exceedingly small value will help us to test an old and celebrated hypothesis that if we could see far enough into space the whole sky would be seen occupied with discs of stars all of perhaps the same brightness as our sun, and that the reason why the whole of the night-sky and day-sky is not as bright as the sun's disc, is that light suffers absorption in traveling through space." Although he coupled together the riddle of cosmic darkness and the absorption solution, and referred to both as "a celebrated hypothesis," he made no mention of Chéseaux and Olbers.

Kelvin argued that the fraction of the sky covered by stellar disks equals the average brightness of the starlit sky at each point compared with the brightness at a point on the Sun's disk:

$$\text{the fraction of sky covered by stars} = \frac{\text{brightness of starlit sky}}{\text{brightness of Sun's disk}}$$

If one-trillionth of the sky is covered with stars, the starlit sky on the average at every point is one-trillionth as bright as the Sun's disk. (In most discussions of Olbers's paradox this equivalence is overlooked.) When the sky becomes fully covered—and the left side of both the

REGION OF NEBULAE

GALAXY OR REGION OF STARS

REGION OF NEBULAE

14.2   Kelvin used Simon Newcomb's book *Popular Astronomy* (1878) containing this diagram of the "probable arrangement of the stars and nebulae." The Galaxy, according to Newcomb, had a radius of 1,000 parsecs and consisted of a billion stars.

above equations then equals unity—the visible universe extends as far as the background limit and the brightness of the sky at every point matches the brightness of the Sun's disk.

On comparing the two equations above, we see that the ratio of the size of the visible universe to the background limit equals, in Kelvin's words, the ratio of "the apparent brightness of our star-lit sky to the brightness of our sun's disc":

$$\frac{\text{size of visible universe}}{\text{background limit}} = \frac{\text{brightness of starlit sky}}{\text{brightness of Sun's disk}}$$

That is, when the size of the visible universe equals the background limit, the starlit night sky at every point is as bright as the Sun's disk. Looking back over more than four hundred years at the numerous proposed solutions of the riddle of cosmic darkness, often presented without any supporting argument and mathematical proof, Kelvin's lucid treatment is a delight to read.

Kelvin's Solution

Kelvin's calculations, based on available astronomical information, showed that the background limit was at a distance of 3,000 trillion light-years. (His result was actually about one-tenth of this value because he allowed for only a partially covered sky.) The background limit was hence a trillion times larger than the Galaxy. If one assumed that no more stars existed outside the Galaxy, then only a trillionth of the sky could be covered with stellar disks, and at every point the sky would be a trillion times less bright than the Sun. This is the old Stoic solution to the riddle of cosmic darkness.

But even if we occupied a star-filled universe or "great sphere" of stars of radius 3,000 trillion light-years, said Kelvin, the night sky still would not be covered with stars because the *visible* universe would have a radius much smaller than 3,000 trillion light-years. Light must take 3,000 trillion years "to travel from the outlying suns of our great sphere to the center" that we occupy. But this vast span of time greatly exceeds even the most liberal estimate of the time that has elapsed since all the stars started to shine. "We have irrefragable dynamics proving that the whole life of our sun as a luminary is a very moderate number of million years, probably less than 50 million, possibly between 50 and 100. To be very liberal, let us give each of our stars a life of a hundred million years." For good reason, as we shall see, he assumed that all stars had been shining for no longer than 100 million years. "Hence," said Kelvin, "if all the stars through our vast sphere commenced shining at the same time . . . at no one instant would light be reaching the earth from more than an excessively small proportion of all the stars."* Lord Kelvin demonstrated with rigor what Edgar Allan Poe had qualitatively anticipated: that the distance to the background is "so immense that no ray from it has yet been able to reach us at all."

"Our supposition of uniform density of distribution is, of course, quite arbitrary," said Kelvin, and we should assume that the density in the great sphere is much less than in the Galaxy. In other words, we should increase the average separating distance between the stars. This would

---

*Entry 9 in the table "Proposed Solutions."

make the background limit even greater, and light would then take even longer to reach us from the outlying stars.

Let us suppose that we live in a universe of unlimited age in which the stars have been shining for only a limited time. In this case, the visible universe (the part we see) cannot be infinite in size. We look out in space to a time when the stars first became luminous. This is the horizon of the visible universe.[7] Beyond the horizon we see the darkness that existed before the birth of stars. We could of course extend the visible universe by arranging for many generations of stars. We would, in Kelvin's treatment, require at least 30 million generations, each lasting 100 million years, for the size of the visible universe to exceed his background limit and cover the sky with stars. But, as Kelvin perhaps realized, this cannot work, because to every luminous star there are now 30 million nonluminous stars, and as Fournier d'Albe had already realized, the sky remains dark because it is covered with nonluminous stars.

Let us now suppose that we live in a universe of finite age. This probably is what Kelvin had in mind, for he appears to have subscribed to a version of the Kant–Laplace nebula hypothesis. In his "Mechanical antecedents" (1854), he wrote, "we know that from age to age the potential energy of the mutual gravitation of those bodies is gradually expended," and in the future "we find that the end of the world as a habitation for man, or for any living creature or plant at present existing in it, is *mechanically inevitable.*" When we trace world history backwards,

> according to the laws of matter and motion, certainly fulfilled in all the action of nature which we have been allowed to observe, we find that a time must have been when the earth, with no sun to illuminate it, the other bodies known as planets, and the countless smaller planetary masses at present seen as the zodiacal light must have been indefinitely remote from one another and from all other solids in space . . . If in purely mechanical science we are ever liable to forget this limitation, we ought to be reminded of it by considering that purely mechanical reasoning shows us a time when the earth must have been tenantless; and teaches us that our bodies, as well as all living plants and animals, and all fossil organic remains, are organized forms of matter to which science can point no antecedent except the Will of a Creator.[8]

In either case—whether the universe is eternal or finite in age—the general condition for a dark starlit sky can be simply stated:

size of the visible universe
must be less than
the background limit.

The Luminous Lifetime of Stars

Kelvin's interest in the problem of determining the age of the Earth and of the Sun lasted throughout his life.[9] At first he supposed that the Sun derived its luminous energy from heat produced by the infall of meteors. In an essay "On the age of the Sun's heat" he wrote in 1862, "The sun must, therefore, either have been created as an active source of heat at some time of not immeasurable antiquity, by an over-ruling decree; or the heat which he has already radiated away, and that which he still possesses, must have been acquired by a natural process, following permanently established laws."[10] Pursuing an idea initially suggested by Julius von Mayer (an innovative German physicist who first proposed the conservation of energy principle in its full generality) and promoted by Hermann von Helmholtz (a versatile German scientist), he supposed that the Sun derived its luminous energy by slow gravitational contraction. He found the Sun's age—derived in this way and now known among astronomers as the Helmholtz–Kelvin or the Kelvin time scale—lay somewhere between 20 and 100 million years. In Kelvin's day, before the discovery of nuclear energy, the release of gravitational energy by slow contraction provided what seemed the most plausible explanation of the Sun's continual loss of radiant energy. Furthermore, it was thought that stars everywhere obtain their energy in this manner.

Stars in an eternal and unchanging universe shine inexhaustibly and the sky blazes unceasingly with brilliant light. This picture presents an unreal world. In the real world, stars have limited reserves of energy; they are born, they shine for a finite lifetime, and then die. We have discovered that the radiation streaming away from stars derives from the subatomic energy released in the furnaces of their deep interiors. Stars of high luminosity consume their nuclear fuel rapidly and shine brightly for millions of years; stars of low luminosity consume their nuclear fuel slowly and shine feebly for hundreds of billions of years. The luminous lifetime of stars of middling luminosity like the Sun is characteristically 10 billion years. The Galaxy, coincidentally, is approximately 10 billion years old. The Sun at present has an age of 5 billion years, and will shine for about another 5 billion years; it is not a first generation star and was born when the Galaxy was already quite old.

Modern astronomy reveals that we live in a big bang type of universe that, so far as we can tell, began about 15 billion years ago. When we take into account the clustering of stars to form galaxies, we find that the background limit is near 100 billion trillion light-years (1 followed by 23 zeros). But the stars that have shone the longest cannot be older than the universe, and probably are no older than the galaxies, and in round numbers have therefore an age of 10 billion years. Just for the purpose of calculation, let us be liberal and assume that all stars are 10 billion years old. Thus, the visible universe has roughly a size of 10 billion light years. With these numbers we find that only one ten-trillionth (10 billion divided by 100 billion trillion) of the sky is covered with stars. Remarkably, this result remains little affected by the fact that the universe is in a state of dynamic expansion.

By knowing how to calculate the background limit, and by varying the size of the visible universe, we can easily design imaginary dark-sky and even bright-sky universes.[11] Our feebly lit sky at night will shine brighter if we bring the stars closer together. To cover the sky with stars, so that each point is as bright as the Sun, we must reduce the background limit to less than 10 billion light-years; stars will then have their average separation reduced to less than one twenty-thousandth of the present value, and will have average separating distances of less than 3,000 times the Sun–Earth distance.[12]

The calculations by Loys de Chéseaux and Wilhelm Olbers yield either explicitly or implicitly a background limit in the region of 1,000 trillion light-years. Possibly, like many other astronomers, they supposed that stars shine inexhaustibly. They certainly assumed that stars shine long enough for light to travel from the background limit. Had they questioned this assumption—that stars, according to their calculations, have been shining for 1,000 trillion years—they might have realized that there was no need to postulate the absorption of starlight.

Kelvin's Bright Sky

We must note that Kelvin solved the riddle of a dark starlit sky according to the conditions prescribed in its original conception. His solution is hence sufficient, for if it can be demonstrated that the sky at night is necessarily dark in a transparent, uniformly populated, and static universe, then all variants of this primitive standard model that resort to absorption, hierarchy, and redshift (owing to expansion) merely accomplish a state of greater darkness in a universe already dark.

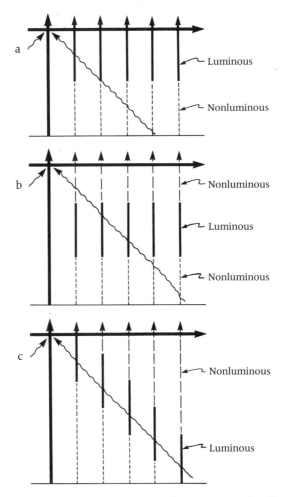

14.3   How stars can be arranged so that an observer sees a bright sky. (a) A Newtonian universe in which stars commence shining everywhere at the same instant. An observer looks out in space beyond the horizon of the visible universe to a time before the stars became luminous. (b) A similar universe in which stars are now all dead. Near the observer the stars are nonluminous, farther away they appear luminous as they were in the past, and still farther away they again appear nonluminous. (c) A universe in which stars commence shining at progressively earlier times as their distances from the observer increase. Kelvin pointed out that with this improbable arrangement the observer sees the sky covered with stars.

Kelvin showed how stars of comparatively short luminous lifetime can be artificially arranged to create a bright sky (Figure 14.3). "To make the whole sky aglow with the light of all the stars at the same time the commencements of the different stars must be timed earlier and earlier for the more and more distant ones," so that all their light arrives at the Earth within the lifetime of the Sun.[13] He pointed out that such an arrangement, which focuses starlight on our part of the universe, would be most improbable.

Kelvin disbelieved in paradoxes. In his *Baltimore Lectures* in 1884 he more than once declared, "There are no paradoxes in science," and in a Royal Institution discourse in 1887 he said the same, "Paradoxes have no place in science." He took the rationalist attitude that paradoxes are the result of misunderstandings; they lie in ourselves and not the external world. It seems historically ironic that he was the first to solve with rigor and utmost lucidity a riddle that later, when his work lay forgotten, became known as Olbers's paradox.

FIFTEEN

Ether Voids, Curved Space, and a Midnight Sun

I have not had a moment's peace or happiness in respect to electromagnetic theory since November 28, 1846. All this time I
have been liable to fits of ether dipsomania, kept away at intervals only by rigorous abstention from thought on the subject.

Lord Kelvin to George FitzGerald, 1896

A CURIOUS solution to the riddle of darkness that received frequent
mention in the astronomical literature of the late nineteenth century
grew out of the old belief that light cannot propagate without an etheric
medium. Lord Kelvin wrestled mightily with the problems of the ether;
in 1846 at age twenty-two he wrote his first paper on electricity and
magnetism, and suffered for the rest of the century from what he called
"ether dipsomania." In 1854 he wrote, "That there must be a medium
forming a continuous material communication throughout space to the
remotest visible body is a fundamental assumption in the undulatory
Theory of Light."[1] The Aristotelian ether had become the luminiferous
medium—the nominative of the verb "to undulate"—updated and
adapted for the transport of light by Hooke, Huygens, Young, and Fresnel, and then mechanized to explain electric and magnetic phenomena
by Michael Faraday, Kelvin, and Maxwell (Figure 15.1).

Ether Voids

Simon Newcomb, a Canadian astronomer and one of the great celestial
mechanicians of the nineteenth century, who became director of the
Naval Observatory in Washington, D.C., discussed in his book *Popular*

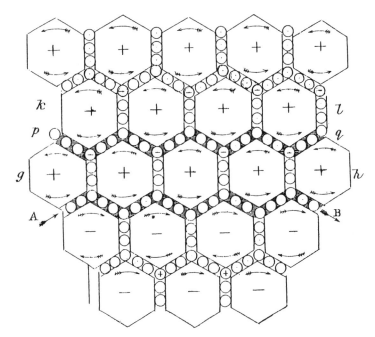

15.1 James Clark Maxwell (1831–1879) visualized a mechanical ether composed of vortices with tiny interstitial vortices acting as "idle wheels." In a paper "On physical lines of force," concerning this model, he wrote in 1861, "It serves to bring out the actual mechanical connexions between the known electromagnetic phenomena; so that I venture to say that any one who undestands the provisional and temporary character of this hypothesis, will find himself rather helped than hindered by it in his search after the true interpretation of the phenomena." A mechanical ether inspired the idea that far away from the Galaxy the ether might vanish, leaving space as an etherless void incapable of transmitting light.

Astronomy (1878) some aspects of the riddle of darkness. Newcomb wrote that the radiant heat from the Sun could conceivably be conserved if the ethereal medium terminated and reflected back the Sun's rays.[2] John Gore elaborated on the idea.

Educated at Trinity College, Dublin, Gore served as an engineer in India for ten years building canals, retired in 1879, and became an astronomer who wrote learned and popular works. In his book *Planetary and Stellar Studies* (1888), he explained that perhaps in the dark abysses of space there exists a total vacuum devoid not only of matter but also of ether. "It has been argued by some astronomers that the number of the stars must be limited, for on the supposition of an infinite number uniformly scattered through space, it would follow that the whole heavens should shine with a uniform light, probably equal to that of the sun."[3] Gore took the general view that the stars and nebulae visible in large telescopes belonged to our own Galaxy; other galaxies may exist, but cannot be seen. Perhaps, he said, following Newcomb's suggestion, these external galaxies are screened from view by an intergalactic absolute vacuum, an etherless gulf, across which waves of light cannot propagate.* All galaxies might thus be screened from one another. What happens to the starlight? "It seems probable that the rays of light from the stars composing our universe would be reflected back from the verge of the vacuum," and therefore "we may consider that the reflecting vacuum as forming the internal surface of a hollow sphere" (Figure 15.2). By universe he meant our Galaxy.

Starlight in the Newcomb–Gore universe cannot cross intergalactic space and consequently each galaxy retains its own light. Apparently neither Newcomb nor Gore realized that the riddle of darkness cannot be solved by surrounding the galaxies with reflecting walls. The starlight trapped in each galaxy bounces to and fro between reflecting walls, and equals more or less the amount of light that would be received from other galaxies in the absence of the reflecting walls.[4] The radiation level rises inside the cavity that surrounds a galaxy, and in the Newcomb–Gore universe we have again a bright sky.

As late as 1902, in an important paper, "On the clustering of gravitational matter in any part of the universe," dealing among other things with the gravitational collapse of sidereal systems, Kelvin wrote, "We need not absolutely exclude, as an idea, the possibility of there being a portion of space occupied by ether beyond which there is absolute vacuum—no ether and no matter."[5] But despite his ether dipsomania, he thought it improbable "that there is a boundary around our universe beyond which there is no ether and no matter."

The argument that the sky should have a uniform brightness equal to that of the Sun at every point depends, wrote Fournier d'Albe in 1907 in the April 12 issue of the weekly *English Mechanic and World of Science,* on the truth of four assumptions:[6]

(1) The stars are irregularly distributed.

(2) The obscuration by dark bodies is negligible.

---

*Entry 10 in the table "Proposed Solutions."

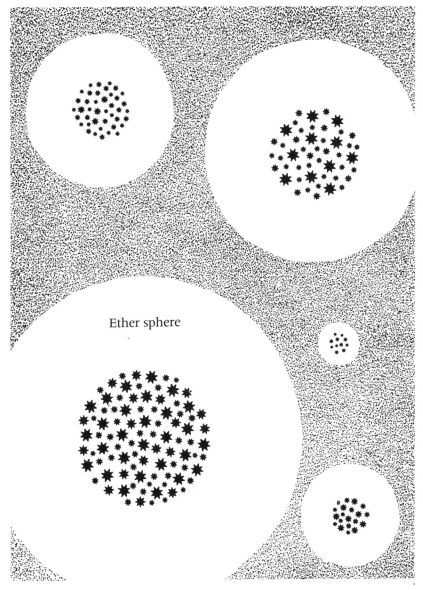

Ether sphere

15.2 The Newcomb–Gore solution of the riddle of cosmic darkness. Each galaxy is immersed in an ether sphere. The intergalactic space between the ether spheres contains no etheric medium and cannot transmit light. In effect, a reflecting surface surrounds each galaxy, and an observer in a galaxy cannot see the other galaxies.

(3)   The luminous stars have eternal existence.

(4)   The luminiferous ether pervades all space.

By "irregularly distributed" he meant scattered in a way that eliminates the possibility of rows of stars hiding behind visible stars as in solution 1 in the table of Proposed Solutions. The absence of obscuration by numerous dark bodies eliminates Fournier d'Albe's solution 7; the eternity of luminous stars eliminates the Poe–Kelvin solution 9; and the pervasiveness of the luminiferous ether eliminates the Newcomb–Gore solution 10; Fournier d'Albe argued that the first, second, and fourth assumptions are plausible (I have actually changed their order), whereas the third is physically impossible.

"Who knows but the interstellar luminiferous ether may thin out and finally forsake us!" On the whole, however, Fournier d'Albe thought this was improbable. The week previously, in the April 5 issue, he rejected the idea of "a gap in the ether all round our stellar system" forming a reflecting surface like a "Ptolemaic firmament." No astronomical evidence indicated reflected starlight, and furthermore such an arrangement has no advantage for it "would preserve the light and heat of our own stellar system from dissipation." He realized that the retained starlight of our own system would compensate for the excluded starlight of other systems.

In the twentieth century we have cast off the habit of thinking in terms of an undulating ether. Einstein's theory of special relativity sealed the doom of this marvelous substance of electric and magnetic stresses, although many scientists at first found it difficult to comprehend that this theory made obsolete an etheric medium for the propagation of light. Nowadays we think in terms of an abstract activity of electromagnetic fields in spacetime.

Curved Space

Albert Einstein, born in 1879 in Germany and educated in Munich and Zurich, became the most outstanding scientist of our time and perhaps all time. Early in this century he brought together the various lines of investigation by many physicists and formulated the theory of special relativity. According to this new theory, everything exists in a universal spacetime, and this common spacetime decomposes into the different spaces and times of things in relative motion. Furthermore, the structure of the four-dimensional continuum of spacetime accounts for the constancy of the speed of light for all observers, demonstrated in the experiments by Albert Michelson and Edward Morley.

A year after becoming professor of physics at the Kaiser Wilhelm Institute in Berlin, Einstein in 1915 brought his theory of general relativity into definitive form in a paper entitled "The equations of the gravitational field." This revolutionary theory showed that gravity is the consequence of the curvature (or geometric deformation) of four-dimensional spacetime, and the curved orbits of bodies in gravitational fields are actually geodesics (straight lines or shortest distances) in curved spacetime. Gravity, a mysterious force acting at a distance, became a property of spacetime whose curvature ripples and variations travel at the speed of light. Space and time, which previously had served as separate and passive frames of the world, became a unified spacetime participating in the activity of the world.

For many decades before the advent of Einstein's theory of general relativity, authors had discussed the possibility of space curvature and considered various experimental methods for verifying its existence. The geometry of a uniformly curved space of finite extent was first studied by the German mathematician Friedrich Riemann in the mid-nineteenth century and later by Simon Newcomb.[7] The surface of a sphere illustrates the properties of spherical space. Each straight line in the surface is a great circle, and all straight lines radiating from any point intersect at the antipodal point on the opposite side of the universe and return to their starting point.

A Midnight Sun

Barrett Frankland, at a meeting in 1913 of the London Branch of the Mathematical Associates, discussed proofs and disproofs of curved space and pointed out that in spherical space we would see the front and back of a star on opposite sides of the sky.[8] "The two images of the star," he said, "seen in opposite directions, appear equally bright if there is no quenching of light in interstellar space." Living in spherical space, we would see two Suns—the real Sun and its antipodal image—equal in size and brightness. Any difference between the two, said Frankland, would come from the fact that we see the Sun in one direction as it was 500 seconds ago and in the other as it was long ago. Darkness vanishes when the Sun shines by day and the antipodal Sun by night (Figure 15.3).

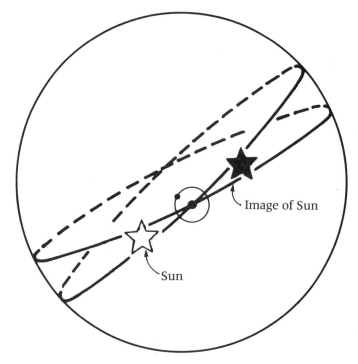

15.3 The midnight Sun. On one side of the Earth lies the Sun, and on the other side lies the antipodal image of the back side of the Sun formed by rays traveling around spherical space.

In 1916 and 1917, in a series of papers "On Einstein's theory of gravitation and its astronomical consequences," the Dutch astronomer Willem de Sitter discussed the dynamic effect of the curvature of space according to the theory of general relativity, and once again raised the problem of darkness in a spherically closed universe. Such a space, he said, "being finite, and the straight line closed, we should, at the point of the heavens opposite the sun, see an image of the back side of the sun. This not being the case, the light must be absorbed on its way 'round the world.'"[9] Apparently de Sitter had not fully considered the consequences of absorption of starlight in interstellar space and the possibility that the absorbing medium would heat up and quickly come into equilibrium with the stars.

In another paper in 1917, "On the curvature of space," de Sitter had a better idea: "It might be argued that we should not see the back of the actual sun but of the sun as it was when the light left it. We could thus do without absorption, if the time taken by light to . . . [circumnavigate the universe] exceeded the age of the sun."[10] Inasmuch as cosmic daylight does not exist, we can regard this as a solution of a variant form of the riddle of darkness.*

A Finite Universe

Johann Zöllner,[11] an ingenious but slightly eccentric German astronomer, claimed in 1883 that a finite, unbounded universe, enclosed within spherical space and containing a finite number of stars, solves the darkness riddle.** This appealing solution, supported by ordinary common sense, has recently been championed by the historian of science Stanley Jaki. At first one feels tempted to think that if the background limit (as determined in flat space) greatly exceeds the circumference of a spherically closed universe, then in a finite universe, much smaller than the background limit, the sky will remain dark at night. Stars will pour forth their rays, and these rays, traveling around the universe and returning to their sources, will not fill space with radiation.

Two comments must be made concerning the Zöllner–Jaki solution. The first is that the curvature of space cannot solve the riddle of cosmic darkness. The random gravitational fields of stars and also of galaxies continually deflect rays of light by small amounts. Hence rays traveling around the universe tend to spread out and not return to their respective sources. Spherical space, instead of acting as a perfect lens, focusing and returning all rays to their sources, acts as an imperfect lens, full of optical aberrations, and beams of light become blurred and defocused. Rays rarely return directly to their sources; constantly they change direction by small random amounts while circulating around the universe, and finally, after many circulations, are intercepted by the surfaces of stars. The average distance traveled by a circulating ray before absorption by a star equals the background limit. The background limit is calculated in exactly the same way as previously, and is the average volume occupied by a star divided by its cross-sectional area. If the distance around the universe is 100 billion light-years, and the background limit is 100 billion trillion light-years, then a ray will circulate on the average a trillion times before absorption.

*Entry 11 in the table "Proposed Solutions."

**Entry 12 in the table "Proposed Solutions."

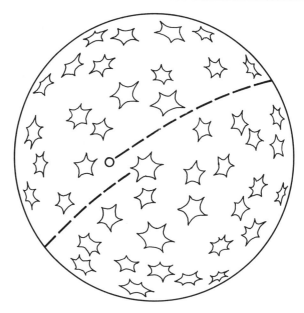

15.4 A line of sight circumnavigates spherical space. This finite space, like the surface of a sphere, is unbounded. Rays of light, as they journey around the universe, suffer multiple small deviations because of the gravitational irregularities produced by astronomical systems. A line of sight, therefore, fails to return to the back of the observer's head; instead, it slightly meanders, and eventually intercepts the surface of a star. The line-of-sight argument used in Euclidean space applies generally to all unbounded and uniform universes, whether finite or infinite.

A line of sight from an observer's eye is a straight line, and in a perfect spherical universe a line of sight, like a straight line, circumnavigates the universe and returns to the observer from the opposite direction. But the gravitational fields of stars and galaxies mar the perfection of spherical space and cause straight lines to deviate. A line of sight, like a beam of light, winds around the universe many times, constantly affected by small random deflections, before intercepting the surface of a star. Every line of sight ultimately reaches the surface of a star, and in a finite closed universe the sky blazes with starlight just as bright as in an infinite open universe (Figure 15.4).

The second point is that in such a universe the sky at night must be dark for exactly the same reason it is dark in a universe of infinite space. If a line of sight, winding round the universe, takes on the average 100 billion trillion years to reach a star, then the sky at night is dark because stars cannot shine for such a long period of time. The Poe–Kelvin solution applies to all unbounded uniform (homogeneous and isotropic) spaces, whether finite or infinite.

In an unclosed universe of infinite extent, uniformly populated with galaxies, stars pour forth their rays, which mingle and fill the whole of space. Similarly, in a spherically closed universe, uniformly populated with galaxies, stars pour forth their rays, which mingle and fill the whole of space. The sky at night is equally bright in the two universes if their stars have similar distributions. Curvature, oddly enough, has no effect on the radiation. This important conclusion is confirmed by powerful thermodynamic arguments, to which we must turn in an attempt to understand what happens in an expanding universe.

The Expanding Universe

But the theory of the expanding universe is in some respects so preposterous that we naturally hesitate to commit ourselves to it. For it contains elements apparently so incredible that I feel almost an indignation that anyone should believe in it—except myself.

Arthur Eddington, *The Expanding Universe*

OF ALL scientific discoveries, the expansion of the universe is the most startling. Nothing in the world around gives warning of the cosmic truth beyond. It came like a thief in the night. Astronomers knew by the mid-1920s that the galaxies were in flight, receding faster the greater their distances from us. By the early 1930s the expanding universe of curved spacetime was widely accepted and securely established.

The first hint of an expanding universe appeared in the work of Vesto Slipher of the Lowell Observatory at Flagstaff, Arizona. For several years, beginning in 1912, Slipher painstakingly determined the velocities of extragalactic nebulae from the shifts in their spectral lines. By 1923 he knew that the majority of the galaxies observed had their spectral lines redshifted—shifted toward the red end of the spectrum—and were therefore receding from our Galaxy.[1]

The second hint came in the series of three papers on the astronomical consequences of Einstein's theory of general relativity communicated in 1916–1917 to the Royal Astronomical Society in London by Willem de Sitter. This work by de Sitter was particularly valuable because at that time Einstein's theory of general relativity was little known outside Germany owing to the 1914–1918 World War between the Allies and Central Powers. Thus, Arthur Eddington, a conscientious objector and director of the Cambridge Observatory, learned of Einstein's theory in its final form and became its principal exponent and champion outside Germany.

Einstein constructed a rather simple model of the universe.[2] He assumed that matter has uniform density in a finite, uniformly curved space, and that furthermore the universe is static, neither collapsing nor expanding. To create a static universe he tamed his dynamic theory by introducing a repulsive force that counterbalanced gravity on the cosmic scale. This strange new force became known as the cosmological term. General relativity theory showed that the universe can be made static provided that space is spherically curved and the cosmological term has a certain precise value. Arthur Eddington would prove many years later that the Einstein static universe exists in a precarious state of unstable equilibrium and small disturbances cause it either to collapse or expand.

Einstein thought that the cosmological term guaranteed a unique global solution. De Sitter in his third paper succeeded in showing that Einstein's universe is not unique, and that it was possible, with the cosmological term, to construct an alternative static model.[3] The odd thing about the de Sitter universe was that it contained no matter. Also it had the peculiar property that when an observer and a particle were inserted, the observer sees the particle moving away. This puzzling behavior of things moving apart became known as the de Sitter effect.

Eddington conjectured that the de Sitter effect had some connection with the recessional motion of the extragalactic nebulae observed by Slipher.[4] Theoretical investigations subsequently demonstrated that the de Sitter universe is not static but actually expanding. The apt remark was then made that the Einstein universe contains matter without motion, and the de Sitter universe contains motion without matter. Together they have the virtue of demonstrating in distinctive forms the properties of spacetime: space in the Einstein universe is static yet curved, whereas space in the de Sitter universe is dynamic yet flat. Generally these properties are combined in states of dynamic curvature.

The deliberate investigation of nonstatic universes began with the Russian physicist Alexander Friedmann in 1922. His work on this subject made little impact, however, until his solutions were rediscovered by the Belgian astronomer Georges Lemaître in 1927. By the early 1930s the expanding universe had become securely established and a totally new understanding of cosmology was in the making.[5]

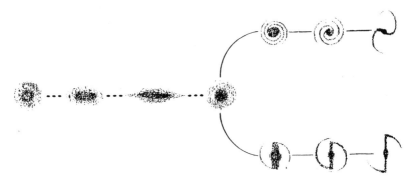

16.1   The Hubble classification of the galaxies. On the left in this tuning-fork diagram are shown the elliptical galaxies arranged in a series of increasing apparent ellipticity, and on the right are shown the spiral galaxies arranged in two parallel series.

The Velocity–Distance Law

Milton Humason used the 100-inch telescope at the Mount Wilson Observatory to extend to greater distances Slipher's measurements of the redshifts of galaxies, and Edwin Hubble, also at the Mount Wilson Observatory, classified galaxies and determined their distances (Figure 16.1). From these early observations and the work of theoreticians emerged the famous velocity–distance relation:

$$\text{velocity of recession} = \text{constant} \times \text{distance}$$

which shows how the velocity of recession of the galaxies increases with their distance (Figure 16.2). Because Hubble played a crucial role in establishing this relation, the "constant" later became known as the Hubble term; its value is still not known with precision.[6]

The assumption that on the average the universe is everywhere the same in space is known as the cosmological principle. This important principle, stating that all places are alike, has become the blueprint of our belief that the universe possesses an underlying homogeneity and exists as a unified whole. It claims that evolving planets, stars, and galaxies are much the same everywhere throughout space, and because they evolve much the same everywhere, the fundamental physical laws are also the same everywhere.

According to the cosmological principle, galaxies recede from one another in similar ways. This fundamental homogeneity explains the

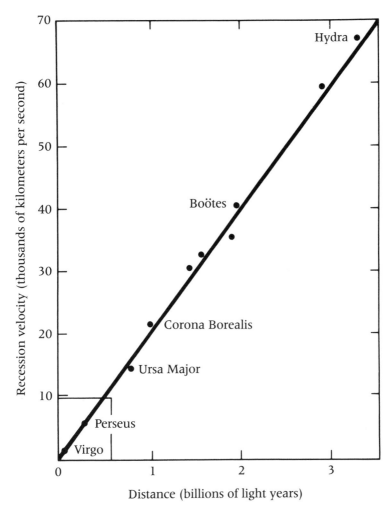

16.2 A representative velocity–distance diagram showing the recession velocity of galaxies increasing with their distance. In the lower left corner the square represents the region surveyed by Hubble up to 1929.

velocity–distance relation. Let us suppose that a galaxy, labeled A, at one billion light-years distance from the Milky Way, recedes from us at roughly one-tenth the speed of light. In the same direction a second galaxy, labeled B, at a distance of one billion light-years from A, recedes from A at the same velocity that A recedes from us. Therefore B, at two billion light-years distance from us, recedes twice as fast, at two-tenths

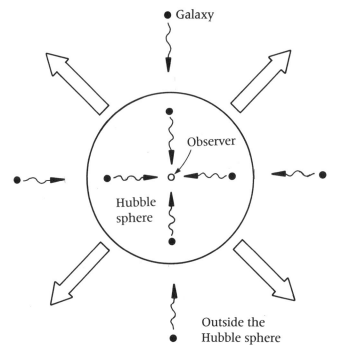

16.3   Inside the Hubble sphere, galaxies recede at less than the speed of light. Outside the Hubble sphere, galaxies recede faster than light, and even the light they emit in our direction recedes as it travels toward us.

the speed of light. Doubling the distance doubles the velocity, and the recession increases linearly with distance because of the underlying homogeneity.

Notice what this means: a galaxy at a trillion light-years recedes 100 times faster than the speed of light! And yet in no circumstances can anything move through space faster than light (Figure 16.3). What happens is that the galaxies, though receding, do not move through space like projectiles; of course they have fidgety (peculiar) motions, and even large-scale streaming motions, caused by their interactions and their tendency to cluster with one another. But these local motions aside, the galaxies sit still in space and let the expansion of space waft them away from one another. Remember, space itself is now dynamic and not just a passive receptacle; and as it expands, the distances between galaxies become greater. (An individual galaxy is held together much too tightly by its own gravity to be pulled apart by cosmic expansion.) The cosmic joyride of the galaxies on expanding space obeys unfamiliar global laws and not the familiar local laws of special relativity.

All distances used in the velocity–distance law must be measured at a common instant in time. And yet we look out in space and back in time and cannot see the galaxies distributed in space at a common instant. We never see them receding exactly in accordance with the velocity–distance law because in the past, at the time they emitted the light that now is reaching us, they were nearer to us than they are at present. In effect, each galaxy has two distances: its present distance, and its distance at the time when it emitted the light that we now see. If we are to compare the recession velocities of galaxies, we must use their present distances in the velocity–distance law, and not their observed distances in the past when they were at different distances. Thus galaxy B, at present at twice the distance of galaxy A, recedes twice as fast; but the observed distance of galaxy B is not twice the observed distance of galaxy A. (In some cases B's observed distance can be less than A's observed distance.)

Also we must be careful and allow for evolution effects in distance estimates that compare galaxies at different distances from us and therefore at different times in the past. And we must allow for the fact that the value of the Hubble term changes with time and was different in the past than at present. Observations of distant galaxies must be adjusted to take these complications into account. The Cartesian fear that a finite speed of light would confuse the appearance of things is fully justified in cosmology.

The First Hundred Thousand Years

Back in the past the universe was denser, and—as first suggested by George Gamow, Ralph Alpher, and Robert Herman, and later confirmed by observations—it was hotter.[7] Our knowledge of cosmic history at the dawn of time contains many gaps; for example, we still do not know how galaxies originated. Nonetheless we can trace in broad outline the threads of cosmic history back to when the universe had an age of one second.

Knowledge of what happened in the extreme early universe depends in part on progress in our understanding of the world of subatomic particles. According to modern theory, many aspects of which no doubt will be out of date in a few years, the universe evolved through successive exotic states of matter during the first second of its existence.

At an age of one second, the universe had a temperature everywhere of 10 billion degrees Kelvin and contained concentrated radiation in the form of x-rays at a mass density one million times greater than that of water. What matter there was, mostly in the form of hydrogen, had roughly the same density as water. The charismatic Gamow and his able young colleagues Alpher and Herman estimated that radiation dominated the early universe for a period of time lasting from about one second to 100 thousand years. They named this period the radiation era.

Throughout the radiation era, when the universe was flooded with intensely bright light, the temperature continued to drop because of cosmic expansion, and the radiation progressively decreased in density. By the end of the radiation era the temperature had dropped to a few thousand degrees Kelvin and was sufficiently low for radiation to decouple from matter. After the radiation era, rays of light moved freely without being constantly scattered by atoms. The densities of matter and radiation continued to decrease, but now the density of matter greatly exceeded the density of radiation.

Expansion of the universe cools the cosmic radiation. As a result, only a very small fraction of the immense energy of the early universe survives in the form of the cool background radiation that now fills the universe and has a temperature of approximately 3 degrees Kelvin (minus 270 degrees Celsius).

Steady-State Universes and the Dark Night Sky

In their paper, "The steady state theory of the expanding universe," the British cosmologists Hermann Bondi and Thomas Gold in 1948 reawakened interest in the riddle of cosmic darkness, which for several years had been overshadowed by the physical and mathematical complexities of an expanding universe of curved spacetime.[8] These authors suggested that explaining the darkness of space is one of the main tasks of cosmology: "The connection between this phenomenon and cosmology was noticed by Olbers, who pointed out that in an infinite homogeneous static Newtonian universe the mean radiation density would be as high as on the surface of a star!" Encumbered with one or two trifling historical inaccuracies, the riddle gained a new lease of life.

A universe in a steady state never changes its appearance; the cosmic scenery, apart from local details, always has been and always will be the same as it is now. The Epicurean and Aristotelian systems, and versions of the static Cartesian and Newtonian systems, were eternally unchanging and therefore in a steady state. These systems were naturally static, neither expanding nor collapsing. William MacMillan, professor of astronomy at the University of Chicago, proposed as recently as 1925 in a three-part paper, "Some mathematical aspects of cosmology," yet another static universe in a steady state.[9] We need not think that the "universe as a whole has ever been or ever will be essentially different from what it is today," wrote MacMillan. The blackness of the night sky meant, he said, that new atoms are "generated in the depths of space through the agency of radiant energy," and hence new matter is constantly created at the expense of radiation.* Stars slowly dissolve into starlight, said MacMillan, which streams away into the depths of space. But instead of accumulating in the universe, creating a bright sky, this radiation slowly transforms back into atoms of matter, which then aggregate to form new luminous stars. Thus, new stars replace old stars, and the new stars in their turn dissolve into radiation, which transforms back into atoms, and so on, in a perpetual cyclic process. In this manner, said Macmillan, energy is conserved and the sky remains forever dark.

This ingenious perpetual-motion universe got a cool reception in scientific circles. Stars cannot radiate away their entire mass, and radiation cannot transform back into matter in the way proposed. Also MacMillan completely ignored entropy in the universe. According to the second law of thermodynamics, entropy must always increase. Hence energy, though conserved, cascades always into less available forms and cannot be perpetually recycled as MacMillan suggested.

Bondi and Gold generalized the cosmological principle—that all places are alike in space—to the perfect cosmological principle: that all places are alike in space and time. They proposed that matter is created continuously throughout space, and the newly created matter aggregates and forms young galaxies, which then occupy the widening gaps between old galaxies. In their universe, energy is not recycled, as in the MacMillan universe. The continuous creation of matter (new energy in a low state of entropy) and the continuous dilution of matter and radiation by expansion maintains the universe in a steady state. The British astronomer Fred Hoyle joined their ranks and the subsequent debate between the members of the rival steady-state and big-bang schools

*Entry 13 in the table "Proposed Solutions."

Big bang

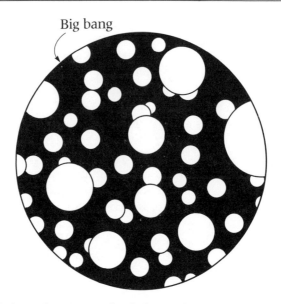

16.4 The big bang, looming in the dark gaps between the stars, covers the entire sky. Once it was incandescent, but now it is invisible because of its extreme redshift. Radiation from the big bang streams in all directions and floods everywhere. Hold up the palm of the hand to the sky, day or night, and 1,000 trillion photons of the big bang will fall on it in one second only.

stimulated considerable interest in cosmology in many sections of society. The debate ended in 1965 with the discovery of the afterglow of the big bang, and the steady-state universe is now mainly of historical interest.

The Horizon and the Big Bang

Twentieth-century discoveries in astronomy have greatly altered our picture of the physical universe. Yet the fundamental idea of a visible universe—the local part of the whole universe that is visible to an observer—remains intact. The visible universe still extends more or less a distance fixed by the flight of light and the luminous lifetime of astronomical systems. How far light travels in the universe depends on the age of the universe, with some corrections because of expansion. From the value of the Hubble term and other observations of the way the universe expands, we know that the universe has an age of roughly 15 billion years.

We look out in space and back in time. How far we look out in space depends on how far we look back in time. At the limit of the visible universe, as far back in time as we can see, lies the horizon. We look back beyond the formation of the earliest stars, beyond the formation of the giant galaxies, to the last moments of the radiation era—at the end of the big bang—when the universe was 100 thousand years old. But the light of the early universe, the big bang, traveling unimpeded from the horizon for roughly 15 billion years, has been greatly cooled and enfeebled by the expansion of the universe (Figure 16.4).

The Cosmic Redshift

On the other hand, the plausible and, in a sense, familiar conception of a universe extending indefinitely in space and time, a
universe vastly greater than the observable region, seems to imply that red-shifts are not primarily velocity-shifts.

Edwin Hubble, *Observational Approach to Cosmology*

WHY IS the sky dark at night? Because, we have been told in recent
years, the universe expands: Starlight from distant galaxies arrives
feeble and red, and the feeblest starlight arriving from the farthest galaxies is redshifted into invisibility. According to this explanation, multitudes of stars actually cover the entire sky, but most of them cannot be
seen because their light is redshifted into invisibility by cosmic expansion. Let us look at this explanation and determine to what extent it is
true.

We know that the universe expands because the light received from
distant galaxies is displaced toward the red end of the spectrum. White
light emitted by stars in galaxies far away and long ago arrives as red
light. As the galaxies drift apart owing to the expansion of intergalactic
space, the waves of light traveling from one galaxy to another are
stretched by the expanding space through which they travel (Figure
17.1). This lengthening of the wavelengths of light shifts the spectral
lines of light toward the red end of the spectrum. The amount of this
redshift (sometimes mistakenly called a velocity or Doppler redshift)
measures the expansion of the universe.[1]

Consider the following: A distant galaxy emits light that astronomers
later detect on Earth. As the light travels through expanding space toward our Galaxy, its wavelengths are steadily stretched. Finally, the light
enters a telescope, and the astronomers compare its spectrum with the
spectra of other sources of light in our Galaxy, and in this way they
measure the amount of redshift. Invoking the cosmological principle,

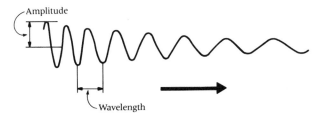

17.1 A wave of radiation is progressively stretched while traveling in expanding space; its wavelength increases and its amplitude decreases.

they assume that the light-emitting atoms in the distant galaxy are similar to the light-emitting atoms in our Galaxy. The amount of redshift detected depends on how much the universe has expanded between the time of emission and the time of reception. We must note that the redshift applies to the whole spectrum, extending from radio waves to visible light to x-rays, and when one wavelength is doubled, all wavelengths are doubled in the spectrum of a source. (The redshift is a percentage sort of measurement: 50 percent increase in the original wavelengths equals a redshift of 1.5; 100 percent increase equals a redshift of 2.)

Let us suppose that the light received from a distant galaxy has its wavelengths increased twofold (equal to a redshift of 1, corresponding to 100 percent increase). From this observation, astronomers immediately know that the universe has expanded twofold since the galaxy emitted the light now arriving. The present distance of the galaxy is therefore twice the original distance at the time of emission. In the same period of time all other extragalactic distances have doubled and the average density of matter in the universe has decreased eightfold. Similarly, if the light received has its wavelengths increased threefold (equal to a redshift of 2, corresponding to a 200 percent increase), the universe has expanded threefold, and the present distance of the galaxy is three times as great as the original distance at the time of emission.

Very remote and highly luminous galaxies have been observed with redshifts as great as 4. The most redshifted light so far detected by physicists and astronomers comes from the big bang. At the end of the radiation era, when the temperature had dropped to around 3,000 degrees Kelvin, radiation decoupled from matter and ever since has traveled freely in the universe, growing progressively cooler. We cannot see this light with the unaided eye because its rays have been redshifted

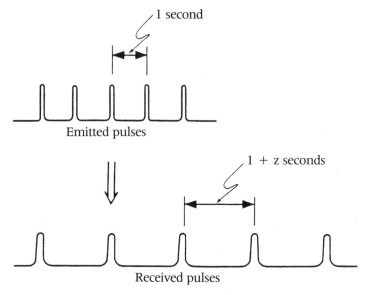

1 second

Emitted pulses

1 + z seconds

Received pulses

17.2 Pulses of radiation are emitted once each second by a galaxy far away and long ago. After traveling through expanding space, the pulses become more widely separated and arrive in our Galaxy at intervals greater than one second. The greater the redshift (denoted by the symbol z) of the radiation we receive from the galaxy, the more widely separated the arriving pulses of radiation. Far away clocks move sluggishly and things appear to age slowly. At the horizon of the visible universe time stands still.

into the far infrared. Arno Penzias and Robert Wilson, using special radio receivers, first detected this radiation in 1965 at the Bell Telephone Laboratories. They found that it has a temperature of approximately 3 degrees Kelvin, not far from the value originally guessed by Gamow and his colleagues. This observation showed that the background radiation from the big bang has now a redshift of 1,000.

The redshift effect applies to all intervals of time. Imagine that a distant galaxy emits a short pulse of light once every second. We must assume (because of the cosmological principle) that the clocks measuring intervals of time in the distant galaxy are similar to the clocks in our Galaxy. The pulses of light, emitted once every second, leave the galaxy with initial separations in space of one light-second. Then, as they travel through expanding space, their separations progressively increase, and eventually the pulses arrive in our Galaxy at a rate slower than one per second because of their wider separations (Figure 17.2). If the wavelengths of the pulsed light are increased twofold while traveling across intergalactic space (equal to a redshift of 1), the pulses will arrive once every two seconds.

We see the things of far away and long ago changing more slowly than local things. The farther away, the slower they change. At the end of the radiation era, revealed by the background radiation, what happened in 1 second is seen by us to happen in 1,000 seconds, more than a quarter of an hour. At the horizon of the visible universe, at the beginning of the big bang, where the cosmic redshift approaches infinity, nothing changes and time stands still. But we cannot see the beginning, only the ending of the big bang at a time when the universe was 100 thousand years old.

The Redshift Solution

During the heyday of the steady-state theory, Hermann Bondi (Figure 17.3) in his publications and lectures explained why a dark night sky is puzzling. He referred to the riddle of darkness as "Olbers' paradox," a title that soon enjoyed wide popularity. In his book *Cosmology,* Bondi wrote in 1952, "If distant stars are receding rapidly the light emitted by them will appear reddened on reception and hence will have lost part of its energy."[2] The starlit sky in an expanding universe is automatically dark, he said, because of the cosmic redshift of light from distant galaxies.*

Bondi listed the principal assumptions in Olbers's treatment of the riddle as seen from a modern viewpoint:[3]

(1) Viewed on a sufficiently large scale the universe is the same everywhere, i.e., it is uniform in space.

(2) The universe is unchanging in time.

(3) There are no major systematic motions in the universe.

(4) The laws of physics, as we know them, apply everywhere throughout the universe.

Olbers explicitly acknowledged assumptions (1) and (4). Assumptions (2) and (3) are implicit in his treatment. He favored, as we know, the idea of a universe containing stars equally spaced or galaxies equally spaced everywhere, either stationary or in orbital motions, all pouring

*Entry 14 in the table "Proposed Solutions."

17.3 Hermann Bondi (b. 1919). With Thomas Gold he collaborated in the origin of the steady-state theory of an expanding universe and revived modern interest in the riddle of cosmic darkness in the context of an expanding universe.

out and filling space with their radiations. A universe unchanging in time is in a steady state; it never evolves, and its stars ceaselessly pour forth radiation into space. Unknown to Olbers, assumption (2) violates the principles of thermodynamics, including the conservation of energy, and therefore it contradicts assumption (4).

Bondi always pointed out in his writings and lectures that Olbers's paradox could be resolved by dropping assumption (2). At the time he made his analysis, however, he was committed to the steady state theory of an expanding universe, which accepted the truth of assumption (2); according to this theory matter and energy are continuously created. He and other steady state theorists argued: not all the assumptions can be correct; granted the truth of assumptions (1), (2), and (4), then it follows that assumption (3) must be false. Olbers's paradox is thus the result, they suggested, of the erroneous assumption that the universe is static. The star-covered sky is dark, and not a blaze of light, because of the expansion of the universe.

Redshift in the Steady-State Universe

In the context of the expanding steady-state universe, proposed by Bondi and Gold, the redshift argument is valid and Bondi was perfectly correct. Although this particular universe is infinite in size and age, it has the peculiar property that the observed part—called the visible universe—has a constant and finite size, equal in this particular universe to the Hubble sphere.

In a *static* steady-state universe, infinite in size and age, the visible universe fills the whole of space. Stars shine endlessly and cover the entire sky with blazing starlight. But in an *expanding* steady-state universe, also infinite in size and age, the visible universe cannot fill the whole of space; for galaxies at vast distances recede faster than the speed of light, and the light they emit, although it hurries toward us through expanding space, itself recedes and never reaches us. The horizon of the visible region in an expanding, steady-state universe lies at the Hubble distance (equal to the speed of light divided by the Hubble term) of approximately 15 billion light-years.[4]

We must also note that in a universe of infinite size and age, an observer's line of sight has infinite length—forgetting for the moment that in a star-filled universe it must terminate at the surface of a star. (Other forms of absorption tend to be irrelevant in solutions of the riddle of darkness.) In the expanding, steady-state universe, what happens is interesting: a line of sight at first extends out in space and also back in time in the normal way; but on approaching the horizon, instead of terminating at the beginning of time, as in the big-bang universe, it bends over and extends back in time, approaching closer and closer the horizon, reaching it in the infinite past. In this ingenious universe, we look out in space only a finite distance and back in time an infinite distance—forgetting, of course, for the moment that at some point a star will get in the way.

Matter is continuously created everywhere in the expanding steady-state universe at a rate of roughly one hydrogen atom per cubic meter every 5 billion years. This matter slowly condenses to form new stars and galaxies. The stars and galaxies born in the visible universe drift toward the horizon, getting redder and redder; ultimately, when they reach the horizon, they have infinite redshift.

In the eternity of the past an infinite number of stars have formed in the visible universe and drifted across the horizon into the invisible regions beyond. An observer's line of sight extending back into the infinite past must ultimately intercept the surface of one of this infinity of stars inside the visible universe. This sounds complicated, but a simple figure shows clearly what happens (Figure 17.4).

In the expanding steady-state universe every line of sight actually intercepts the surface of a star; the sky is covered with stars, and the first interpretation of the riddle of darkness is correct. The golden walls of Edgar Allan Poe are at a distance of about 15 billion light-years (the Hubble distance), consisting of 100 billion trillion trillion stars, most of which were shining 100 billion trillion years ago. But, as Bondi explained, we cannot see this background of glaring stars because most stars composing the background, crowded close to the horizon, have the enormous redshift of 10 trillion.

Doubts about the Redshift Solution

When it was first proposed by Bondi, the redshift solution to the riddle of darkness, valid in the steady-state theory of continuous creation, seemed to many scientists sufficiently plausible to be accepted without question as the general solution to the riddle of cosmic darkness, even in universes of finite age, such as the widely accepted big-bang universe.

Go out at night, astronomers urged members of their audience, and look up at the dark starlit sky. Although countless stars cover the sky, you will see relatively few because most are reddened into invisibility by the expansion of the universe. The darkness of the night sky proves that the universe is expanding. Here was a theme, prefaced with a few words about the Doppler effect, that captured the imagination of a wide audience. Some astronomers went so far as to claim that expansion of the universe was the necessary as well as the sufficient condition for a state of darkness at night.

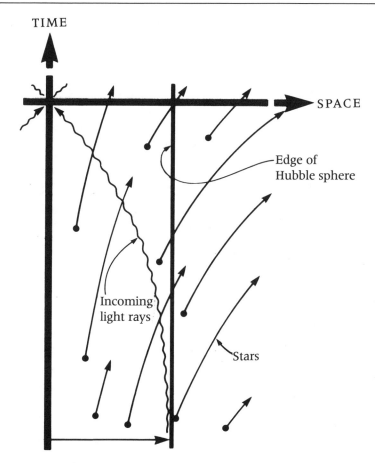

17.4 In a steady-state universe the observer cannot see events outside the Hubble sphere, and most extragalactic light reaching the observer comes from sources close to the edge of the Hubble sphere. In this universe the sky is covered with stars, as Hermann Bondi said, but the starlight is redshifted into invisibility. This explanation fails, however, in the conventional big-bang universe.

I became interested in the subject of Olbers's paradox in the early 1960s on reading Bondi's celebrated book. Like many other investigators, I wrote down a cumbersome mathematical expression that integrated the contributions of radiation from sources distributed in uniformly curved space. The expression involved a double integration over the observer's backward light cone, taking into account absorption and redshift.[5] I stared at it and managed to convince myself that the curvature of space was indeed an irrelevant issue, as Bondi had said. Obviously the redshift effect and dilution by expansion reduced the radiation intensity, and I accepted this as evidence of the general validity of Bondi's solution to the riddle of darkness.

As a parting gesture before turning to other interests, I estimated the amount of energy that would be needed to create a bright sky of the kind imagined by Wilhelm Olbers. The result at first seemed unbelievable. Then it dawned on me that something was seriously wrong with the way we were looking at the whole problem. My idle curiosity in Olbers's paradox flared up into a strong interest that has lasted ever since. It became a hobby on which I would spend a few days a year. I must admit, I still do not know all the answers.

At first I was overly critical of Bondi's redshift solution. (It is possible, for example, to design an expanding steady-state universe in which the sky at night is bright.) While others upheld it as the solution for all universes, I condemned it as the solution for none. Some years passed before I appreciated the full subtlety of the expanding, steady-state universe and realized that in the context of this particular universe, Bondi's redshift argument remains perfectly correct, provided the background limit exceeds the Hubble distance. The first interpretation of the dark night sky applies in the steady-state universe: The whole sky is covered with stars, but the starlight is not perceptible because of its redshift. The golden walls are there, not far away, just redshifted into invisibility.

But our world is now known to be an evolving big-bang universe, and not a steady-state universe of continuous creation. The energy argument is hence valid, and the second interpretation applies: the sky is dark at night because of missing stars and not because of missing starlight.

Energy in the Universe

For in and out, above, about, below
'Tis nothing but a Magic *Shadow*-show
Played in a Box whose candle is the Sun
Round which we Phantom Figures come and go.

The Rubáiyát of Omar Khayyám

ALBERT EINSTEIN, in his essay "$E = Mc^2$," wrote, "To understand the law of the equivalence of mass and energy we must go back to two conservation or 'balance' principles which, independent of each other, held a high place in pre-relativity physics."[1]

The law of matter conservation, accepted in the seventeenth century, served until the beginning of the twentieth century to define what constitutes the material world. Matter, as measured by weight, remained constant during its chemical and physical transformations.

The law of energy conservation followed much more slowly. Until the nineteenth century the flow of heat seemed to imply the continuity of an imponderable caloric fluid, and the interchange of kinetic and potential energies in swinging pendulums and in waterfalls seemed to imply the continuity of mechanical energy. Yet heat and mechanical energy were undoubtedly related by the generation of the former at the expense of the latter in friction, and by the generation of the latter at the expense of the former in steam and gas engines. Benjamin Thompson (Count Rumford) reported to the Royal Society in 1798 his measurements of heat generated by mechanical work in a paper entitled, "Experimental inquiry concerning the source of heat excited by friction," casting doubt on the caloric theory. James Joule reported to the same society in 1849 a more precise relation in his paper "On the mechanical equivalent of heat." In the meantime, scientists such as William Thomson (later Lord Kelvin) had developed an improved understanding of temperature. The principle of the perpetuity or conservation of energy in the form of the first law of thermodynamics, promoted by Joule, proposed in full generality by Julius Mayer, and advanced by Hermann von Helmholtz, was firmly established by mid-nineteenth century.

Although mechanical energy can be entirely dissipated in heat, not all heat can be converted into mechanical energy. Energy in an isolated system remains constant, said Kelvin, but rearranges into progressively less available forms and finally consists of heat at uniform temperature. On the basis of Sadi Carnot's cyclic theory of heat engines, Kelvin and Rudolf Clausius formulated the second law of thermodynamics. In the new and powerful theory of thermodynamics, Kelvin's "unavailability of energy" became Clausius's entropy. Entropy always increases, never decreases, and energy, though conserved, as a consequence tends to rearrange into less available forms.

The theory of relativity demonstrates that all forms of energy—mechanical, thermal, radiant, electrical, atomic, and subatomic—have mass. Thus $E = Mc^2$, or energy equals mass times the square of the speed of light. Alternatively, $M = E/c^2$, or mass equals energy divided by the square of the speed of light. The electrical energy generated each day by a 1,000 megawatt power station has a mass of one gram; this mass of energy flows out of the generators along the transmission lines. The Sun's radiation incident each second on the Earth's surface has a mass of 2 kilograms.

The old law of conserved matter has been discarded and replaced by the more general law of conserved mass that embodies the conservation of energy. Matter no longer is fully conserved in chemical reactions that release and absorb heat. The heat released by coal when burned has a mass of about one ten-billionth that of the coal.

Not Enough Energy

To see what is at issue in the riddle of darkness, let us perform a thought experiment using the equivalence of mass and energy. Let us annihilate all matter in its various forms everywhere in the universe and convert it into thermal radiation. Astronomers estimate from the masses and distribution of galaxies that the average amount of matter in the universe is equivalent roughly to one hydrogen atom per cubic meter. We take one hydrogen atom and convert its mass into thermal radiation occupying one cubic meter. This gives us the same result as the annihilation of all matter everywhere.

To our astonishment we find that the radiation has a temperature of about only 20 degrees Kelvin (20 degrees above absolute zero, or minus 253 degrees Celsius).[2] This temperature is very much less than the 6,000 degrees Kelvin at the surface of the Sun. Thus, all the energy in the universe in the form of matter, when transformed into thermal radiation, still falls a long way short of creating the intensely bright starlit sky feared by Halley, Chéseaux, Olbers, and many other astronomers. Independently of whether the universe is expanding, contracting, or remaining static, a bright starlit sky requires 10 billion times more energy than can be obtained by the drastic measure of converting all matter into radiation.

We have found that the annihilation of all matter yields radiation at a temperature of 20 degrees Kelvin. To create a bright starlit sky therefore requires more than 10 billion stars for every star now shining. During their luminous lifetime, however, stars convert only about one-thousandth of their total mass into luminous energy. More realistically, we therefore find that a bright starlit sky requires at least 10 trillion stars for every star now shining.

The energy solution* shows that there is not enough energy in the universe to create a bright starlit sky. This solution overrides all other proposed solutions. If the universe lacks the energy necessary to create Poe's golden walls, then obviously all the arguments involving other considerations are of secondary importance. In no circumstances can the sky be bright with starlight in a universe such as ours. The stars are too far apart from one another to fill the intervening spaces with thermal radiation up to the temperature of their surfaces.

The elegant differential equations of thermodynamics, when applied to the study of radiation in the universe, are much more enlightening than cumbersome time-retarded integral equations in curved and expanding space. The cosmic box method of studying a sample of the universe in a cavity of varying volume makes immediately apparent what happens in expanding, static, or contracting universes.[3]

A Box Whose Candle Is the Sun

Consider first a static universe of the kind in which the riddle of darkness was conceived. Most of the stars that supposedly cover the sky lie near the background limit of 100 billion trillion light-years. But stars shine typically for 10 billion years and we see them stretching away to

---

*Entry 15 in the table "Proposed Solutions."

a distance no greater than 10 billion light-years. At greater distances, beyond the horizon, we look back over periods of time greater than the luminous lifetime of stars. This picture explains why the sky is dark at night, but how does it relate to the energy argument?

Let us now make a dramatic change in this picture, and imagine each star is surrounded with perfectly reflecting walls. The clustering of stars into galaxies can make little difference in a universe where the sky blazes with intense starlight. We might just as well assume that all stars are uniformly separated, with each occupying its own large average volume. We now suppose that the universe is divided with partitions into cells of equal volume, and that each cell contains one star. The starlight that normally escapes from each star into space now remains confined by the reflecting walls. Instead of streaming away into endless space, mingling with the light from other stars, the light from each star bounces from wall to wall trapped inside a box. Intuitively we realize that the perfectly reflecting partitions have changed nothing; the radiation remains everywhere the same with or without the partitions. Let us remove the partitions, leaving only one typical star enclosed within a cosmic box (Figure 18.1). The star inside the box retains its radiation and fills its own region of space, whereas the many stars outside the box mingle their radiations and fill the universe, and in both cases the radiation is identical.

Imagine that the sky blazes at every point with intense starlight. Wherever we stand in the universe we are surrounded by Poe's golden walls shining with blinding light. We stand, in effect, inside a white-hot furnace, and space is filled with radiation up to the level where it equals the intensity at the surfaces of stars. The time taken by all stars to fill all space with radiation up to this level equals the time taken by one star to fill its own space in the cosmic box with its own radiation.

Rays of light from the star in the box bounce to and fro between the perfectly reflecting walls and are eventually intercepted and absorbed by the star itself. The average time between the emission and absorption of the rays equals the fill-up time of the box. The star then absorbs as much radiation as it emits and the box is filled with radiation in equilibrium with the surface of the star.

Unconfined rays of starlight outside the box travel on the average a distance equal to the background limit before interception by stars. Intuitively we realize that confined rays of starlight inside the perfectly reflecting box travel on the average a distance also equal to the background limit before interception by the star. Folding up the path of a ray

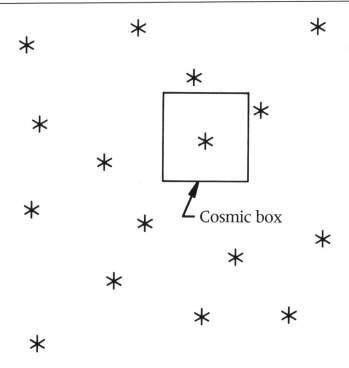

18.1 Imagine a star surrounded by perfectly reflecting walls that form a box having the same volume as the average volume of space occupied by each star in the universe. The conditions for filling the box with radiation from the star are the same as the conditions for filling the universe with radiation from all stars. This is true in static, expanding, and contracting universes. (E. R. Harrison, *American Journal of Physics* 45:123)

of light and putting it inside the box does not alter the length of the path. Emission and absorption by the same star inside the box is equivalent to emission and absorption by different stars outside the box. The multiple images of one star seen in the reflecting walls mimic the appearance of the outside universe of many stars. The cosmic box is a looking-glass universe. Analogously, a tree surrounded by mirrors is a looking-glass forest.

The cosmic box is sufficiently small that we do not have to bother about the cosmic curvature of space; also we do not have to integrate the contributions from numerous stars distributed over the whole of space. The picture of what happens in the cosmic box is simple and the calculations are straightforward.

The fill-up time of the cosmic box is the time taken by light to travel a distance equal to the background limit. This is also the time taken to fill the universe with radiation. Measured in years, it is 100 billion trillion or 1 followed by 23 zeros. But a typical star shines for only 10 billion years (1 followed by 10 zeros), and therefore the box, and hence the universe, cannot fill with radiation in the lifetime of a star, or the lifetime of a galaxy, or the lifetime of the universe. The horizon at a distance of about 10 billion light-years falls short of the background limit by a factor of 10 trillion (Figure 18.2).

In 1901, using slightly different figures, and a different argument, Kelvin showed that the fraction of the sky covered by stars equals the ratio of the radiation density in space to the radiation density at the surface of a star. Our simple picture helps to show why dark gaps must exist between the stars in a universe containing insufficient energy to create a bright sky.

An Expanding Cosmic Box

We now consider an expanding universe. Let us suppose that the cosmic box with its single star expands with the universe. At each instant the volume of the box equals the average volume occupied by a typical star outside the box. Light inside the box bounces to and fro between the slowly receding walls and repeatedly receives small Doppler redshifts. It was shown by Max Planck in 1913 in *The Theory of Heat Radiation* that the addition of many incremental Doppler redshifts reproduces what we call the cosmic redshift (the continuous stretching of wavelengths), and the radiation inside the cosmic box remains identical with the radiation in the universe outside.

If the sky at night is dark because the universe expands, then the radiation in the box is feeble because the box expands. But calculations show that generally starlight in an expanding box is not greatly weaker than starlight in a static box. Most of the radiant energy consists of recently emitted rays of small redshift. The average redshift of all rays is modest and in most models less than unity.[4] The redshift argument claims that the intensity of starlight is reduced to one ten-trillionth of its value in a static universe. The cosmic box shows, however, that expansion cannot enfeeble starlight by this enormous amount (Figure 18.3). As we have seen, the radiation is already weak in a static box and nothing can be gained by expansion except further weakening.

With the cosmic-box method we can design evolving and steady-state

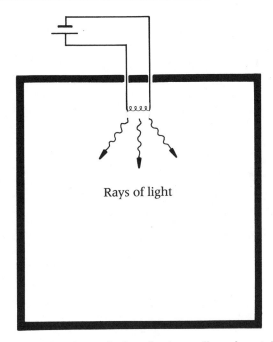

Rays of light

18.2   Imagine a box having perfectly reflecting walls and containing a small source such as a flashlight filament; the filament is the only object in the box that emits and absorbs light. Emitted rays of light travel to and fro between the walls and finally are absorbed by the filament. The average distance traveled by a ray between emission and absorption is the background limit, equal to the volume of the box divided by the effective area of the filament. If the sides of the box have length 1 kilometer and the filament an effective area 1 square millimeter, the average distance traveled by a ray is 1 trillion kilometers, or one-tenth of a light-year. After about five weeks of continuous emission, the filament absorbs as much radiation as it emits. Hence, the box takes five weeks to fill with thermal radiation. At every point in the box the radiation is then as fiercely bright as at the surface of the filament; in all directions the walls glare as bright as the filament itself. But suppose the source of energy—a flashlight battery—can keep the filament bright for a comparatively short time only. In that case the box never fills with radiation because the available energy is insufficient. Analogously, stars contain insufficient energy to fill the universe with radiation in equilibrium with their hot surfaces.

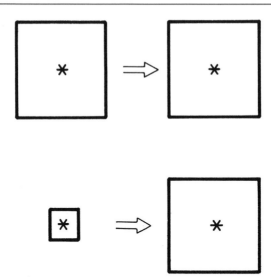

18.3 The redshift of radiation makes the sky at night dark in an expanding steady-state universe but not a big-bang universe. The redshift solution, popular for many years, assumes that the sky is covered with stars and that most stars cannot be seen because of their extreme redshift. Imagine two identical stars, which shine for the same period of time and are confined in separate cosmic boxes, as shown on the left of the diagram. The first box is static and the second box is smaller and expands with the universe. Consider now a later time, shown on the right of the diagram, when the small box has expanded and its volume equals that of the static box. It can easily be shown that the radiation energy in the expanding box is generally never less than one-half the radiation energy in the static box. This demonstrates that expansion cannot generally explain why the sky seems dark.

universes—expanding, static, and contracting—to have either dark or bright skies by simply adjusting the available energy. This may be done by adjusting the average separating distance between stars; thus, the closer they huddle together, the smaller is the volume of the box and the shorter becomes the fill-up time; or equivalently, the closer they huddle together, the closer becomes the background limit and the more the sky is covered with stars.

A sky covered with stars in our universe as constituted at present contradicts the conservation of energy. The expanding, steady-state universe with its continuous creation of matter (and hence energy) does not obey this natural law, and this explains why the energy solution fails for this one particular universe. In a region of space of the expanding, steady-state universe, having a volume say of a million cubic light-years, an infinite number of generations of stars has previously existed. They have all drifted away, and only the expansion of the universe has maintained matter and radiation at a constant level and prevented both from accumulating without limit.

The Elusive Riddle

The energy argument indicates that the riddle of the dark night sky can be solved in an unbounded universe of the kind considered by Digges and Bruno. Of course thermodynamics and relativity theory are recent developments and could not have been used in early investigations. But we have only to compare the solutions of Poe and Kelvin with the energy solution to see that the problem of darkness can be approached and even understood by more than one method. The finite speed of light discovered in the seventeenth century leads to a valid solution when combined (1) with the determination made at about the same time of the separating distance of stars and (2) with the Judaic-Islamic-Christian belief that the universe is of finite age. These considerations lead to the conclusion that the visible universe is finite and the visible stars are too few to cover the sky. The stars needed to cover the sky cannot be seen because their light has yet to reach us. The riddle in fact is solved in its original context by any argument showing that the visible starlit region of the universe extends to a distance less than the background limit.

Absorption, hierarchy, expansion, and any other modification of a static uniform universe like the one in which the riddle was conceived cannot be the answer. A static universe of widely separated stars of finite age is already dark because of the finite speed of light. We can show with modern arguments that these considerations lead to the conclusion that the universe as constituted at present lacks sufficient energy to create a bright sky. Most versions of the absorption solution patently violate the conservation of energy. All versions of the hierarchy solution require clustering out to and beyond the background limit and therefore also violate conservation of energy, because stars do not shine for billions of trillions of years. Expansion also cannot be the answer except in a universe, such as the steady-state universe, that violates the conservation of energy.

One might with justice claim that the age-old riddle of darkness is solved, and all that remains is the question of why most persons who gave the riddle attention thought they could afford to ignore the speed of light. The answer, to me at least, is not at all obvious, and I can only offer one or two guesses.

From the time of Edmund Halley it has been customary to construct concentric spherical shells stacked one on top of the other until stars cover the sky geometrically. This god's-eye view of stars arranged in space without thought of time may have encouraged astronomers to ignore the transit time of starlight.

On the other hand, without thought, we assume that a line of sight in an unbounded universe extends to infinity like a visual ray of the ancient world traveling at infinite speed. Undoubtedly this primitive and automatic way of thinking about vision contributed its mite to the confusion.

The idea of a finite visible universe surrounding an observer like a pool of light, beyond which extends an unlimited invisible universe, emerged slowly and received cautious assent. Almost any estimate of the luminous lifetime of stars, such as a few thousand years according to the Mosaic chronology, or Newton's 50 thousand years for a cooling globe of iron, or 100 thousand years from Kant's cosmogony and Buffon's theory of a cooling Earth, or Kelvin's 100 million years, or any other value less than 100 billion trillion years, would have shown that the visible universe contains far too few stars to cover the sky. Little thought is needed on our part and presumably was needed on the part of astronomers in the nineteenth century to realize that at the horizon of the visible universe exists the creation—either the beginning of the first luminous stars or the beginning of the whole universe. Possibly an understandable reluctance to face the awesome fact that the heavens disclose the creation contributed also its mite to the confusion. It was not the most suitable topic of conversation in Victorian drawing rooms.

But all this cannot be the whole answer because it fails to account for the elusiveness of the solution in the twentieth century. Here we must admit that the complications introduced by an expanding universe contributed more than just a mite to the confusion and perplexity.

Conceivably many astronomers and other persons in the nineteenth century suspected the true answer. Some, including the radical Edgar Allan Poe, alluded to the possibility that starlight from distant stars has not yet reached the Earth. But only Lord Kelvin in a soon-forgotten and little-known paper performed the calculations showing that the sky is necessarily dark in a universe constituted as our own.

Epilogue

ASTRONOMERS in the twentieth century, exploring the cosmic depths, have discovered multitudinous galactic systems. At vast distances they see the bright quasars whose light has taken billions of years to reach us on Earth. From vaster distances, close to the horizon of the visible universe, come numerous but feeble rays whispering the inchoate message of creation.

How we interpret the "darkness behind the stars," in Thomas Dick's words, depends on the nature of the imagined universe in which we live. In the Aristotelian system the dark gaps revealed the outer boundary of the celestial spheres; in the Stoic system they revealed the emptiness of the extracosmic void; in the static star-filled Newtonian system they revealed the prenatal nothingness that existed before the birth of stars. But what do they reveal to the eyes of the modern observer?

Out of doors at night we gaze up at the dark sky. Between the stars we look out to immense distances in space and far back in time before the formation of the galaxies and their firstborn stars. Our vision extends to the horizon of the visible universe, at the frontier of the big bang. In all directions, everywhere between the stars, we see the creation of the universe covering the entire sky. Long ago, when the universe was young and brimmed with energy, the primordial heavens blazed with fierce light. The once-brilliant light of the early universe has gone, cooled a thousandfold by cosmic expansion and transformed into an infrared gloom invisible to the naked eye. In one sense the golden walls of Edgar Allan Poe exist—but mercifully veiled from view. Though we see only a wall of darkness, the big bang covers the sky, filling the universe throughout space and time with its afterglow.

PROPOSED   SOLUTIONS

APPENDIXES

NOTES

BIBLIOGRAPHY

INDEX

Proposed Solutions to the Riddle of Darkness at Night

	Solution	Interpretation	Author	Chapter
1	Stars hidden in rows	B	Fournier d'Albe	1
2	Starlight too feeble	A	Digges	3
3	Dark cosmic wall	B	Kepler	4
4	Stoic extracosmic void	B	Guericke	5
5	Geometric effect	A	Halley	6
6	Interstellar absorption	A	Chéseaux and Olbers	8
7	Obscuration by dark stars	A	Fournier d'Albe	8
8	Hierarchical structure	B	Herschel and Proctor	11
9	Stars not old enough	B	Poe and Kelvin	13, 14
10	Etherless intergalactic space	B	Newcomb and Gore	15
11	Sun not old enough	C	de Sitter	15
12	Finite unbounded space	B	Zöllner and Jaki	15
13	Steady-state universe	A	Macmillan	16
14	Redshift	A	Bondi	17
15	Insufficient energy	B	Harrison	18

A. The covered-sky interpretation.
B. The uncovered-sky interpretation.
C. The midnight-sun interpretation.

Digges on the Infinity of the Universe

A PERFIT DESCRIPTION OF THE CÆLESTIALL
Orbes according to the most aunciente doctrine of the
*PYTHAGOREANS, latelye reuiued by COPERNICVS
and by Geometricall Demonstrations approued*[1]

Althoughe in this most excellent and dyffycile parte of Philosophye in all times haue bin sondry opiniõs touchĩg the situation and mouing of the bodies Celestiall, yet in certaine principles all Philosophers of any accompte, of al ages haue agreed and cõsented. First that the Orbe of the fixed starres is of al other the moste high, the fardest distante, and comprehendeth all the other spheres of wandringe starres. And of these strayinge bodyes called *Planets* the old philosophers thought it a good grounde in reason yᵗ the nighest to the center shoulde swyftlyest mooue, because the circle was least and thereby the sooner ouerpassed and the farder distant the more slowelye . . .[2]

The first and highest of all is the immoueable sphere of fixed starres conteininge it self and all the Rest, and therefore fyxed: as the place vniuersal of Rest, whereunto the motions and positions of all inferiour spheres are to be compared. For albeit sundry Astrologians findinge alteration in the declination and Longitude of starres, haue thought that the same also shoulde haue his motion peculiare: Yet *Copernicus* by the motions of the earth salueth al, and vtterly cutteth of the ninth and tenth spheres, whyche contrarye to all sence the maynteyners of the earthes stability haue bin compelled to imagine.

The first of the moueable Orbes is that of Saturn, whiche beinge of all other next vnto that infinite Orbe immoueable garnished with lights innumerable is also in his course most slow, & once only in thirty years passeth his *Periode.*

The second in Jupiter, who in .I2. yeares perfourmeth his circuit.

Mars in .2. yeares runneth his circulare race.

Then followeth the great Orbe wherein the globe of mortalitye inclosed in the Moones Orbe as an *Epicicle* and holdynge the earth as a Centre by his owne waight restinge alway permanente, in the middest of the ayre is caryed rounde once in a yeare.

In the fift place is Venus makinge her reuolution in .9. moneths.

In the .6. is Mercury who passeth his circate in .80. dayes

In the myddest of all is the Sunne.

For in so stately a temple as this who woulde desyre to set hys lampe in any other better or more conuenient place then thys, from whence vniformely it might dystribute light to al, for not vnfitly it is of some called the lampe or lighte of the worlde, of others the mynde, of others the Ruler of the worlde.

Ad cuius numeros & dij moueantur & Orbes
Accipiant leges præscriptaque fædera seruent.[3]

Trismegistus calleth hym the visible god. Thus doth the Sun like a king sitting in his thrōe gouern his courts of inferiour powers: Neither is yᵉ Earthe defrauded of the seruice of yᵉ Moone, but as *Aristotle* saith of all other the Moone with the earth hath nighest alliance, so heere are they matched accordingely.

In this fourme of Frame may we behould sutch a wonderful *Symetry* of motions and situations, as in no other can bee proposed: The times whereby we the Inhabitauntes of the earth are directed, are constituted by the reuolutions of the earth, yᵉ circulation of her Centre causeth the yeare, the conuersion of her circumference maketh the naturall day, and the reuolutiō of the Moon produceth the monethe. By the onelye viewe of thys *Theoricke* the cause & reason is apparante why in Jupiter the progressions and *Retrogradations* are greater then in Saturn, and lesse then in Mars, why also in Venus they are more then in Mercury. And why sutch chandges from *Direct* to *Retrograde Stationarie. &c.* happeneth notwithstandinge more rifely in Saturn then in Jupiter & yet more rarely in Mars, why in Venus not so cōmonly as in Mercury. Also whye Saturn Jupiter and Mars are nigher the earth in their *Acronicall* then in their *Cosmicall* or *Heliacall* rysinge. Especially Mars who rising at the Sunne set, sheweth in his ruddy fiery coollour equall in quantity with *Iupiter,* and contrarywise setting little after the Sunne, is scarcely to be discerned from a Starre of the seconde light. All whiche alterations appararantlye folowe vppon the Earthes motion. And that none of these do happen in the fixed starres, yt playnly argueth their huge dystance and inmēsurable Altitude, in respect whereof this great Orbe wherein the earth is carryed is but a poyncte, and vtterly without sensible proportion beinge compared to that heauen. For as it is perspectiue demonstrate, Euery quantity hath a certaine proportionable distance whereunto yt may be discerned, and beyond the same it may not be seene, this distance therefore of that immoueable heauen is so exceadinge great, that the whole *Orbis magnus* vanisheth awaye, yf it be conferred to that heauen.

Heerein can wee neuer sufficiently admire thys wonderfull & incomprehensible huge frame of goddes woorke proponed to our senses, seinge fyrst thys baull of yᵉ earth wherein we moue, to the common sorte seemeth greate, and yet in* respecte of the Moones Orbe is very small, but compared with *Orbis magnus* wherein it is caried, it scarcely retayneth any sensible proportion, so merueilously is that Orbe of Annuall motion greater then this litle darcke starre wherein we liue. But that *Orbis magnus* beinge as is before declared but as a poynct in respect of the immēsity of that immoueable heauen, we may easily consider

*Text reads *andte y in*—obviously a printer's error.

what litle portion of gods frame, our Elementare corruptible worlde is, but neuer sufficiently be able to admire the immensity of the Rest. Especially of that fixed Orbe garnished with lightes innumerable and reachinge vp in *Sphæricall altitude* without ende. Of whiche lightes Celestiall it is to bee thoughte that we onely beholde sutch as are in the inferioure partes of the same Orbe, and as they are hygher, so seeme they of lesse and lesser quantity, euen tyll our sighte beinge not able farder to reache or conceyue, the greatest part rest by reason of their wonderfull distance inuisible vnto vs. And this may wel be thought of vs to be the gloriouse court of yᵉ great god, whose vnsercheable worcks inuisible we may partly by these his visible cōiecture, to whose infinit power and maiesty such an infinit place surmountinge all other both in quantity and quality only is conueniente. But because the world hath so longe a tyme bin carryed with an opinion of the earths stabilitye, as the contrary cannot but be nowe very imperswasible, I haue thought good out of *Copernicus* also to geue a taste of the reasons philosophicall alledged for the earthes stabilitye, and their solutions, that sutch as are not able with *Geometricall* eyes to beehoulde the secret perfection of *Copernicus Theoricke*, maye yet by these familiar, naturall reasons be induced to serche farther, and not rashly to condempne for phantasticall, so auncient doctrine reuiued, and by *Copernicus* so demonstratiuely approued.

VVhat reasons moued Aristotle and others that folovved him to thincke the earth to rest immoueable as a Centre to the vvhole vvorlde.

The most effectuall reasons that they produce to proue the earthes stability in the middle or lowest part of the world, is that of Grauitye and Leuitye. For of all other the Elemente of the earth say they is most heauy, and all ponderous thinges are carryed vnto it, stryuinge as it were to sway euen downe to the inmoste parte thereof. For the earth beinge rounde into the which all waighty thinges on euery side fall, makinge ryghte angles on the superficies, muste neades if they were not stayde on the superficies passe to the Center, seinge euery right line yᵗ falleth perpendicularly vpon the *Horizon* in that place where it toucheth the earth muste neades passe by the Centre. And those thinges that are carried towarde that *Medium,* it is likely that there also they woulde reste. So mutche therefore, the rather shall the Earth rest in the middle, and (receyuinge all thinges into yt selfe that fall) by hys owne wayghte shall be moste immoueable: Agayne they seeke to proue it by reason of motion and his nature, for of one and the same a simple body the motion must also be simple saith *Aristotle.* Of simple motions there are two kyndes right and circulare, Right are either vp or downe: so that euery simple motion is eyther downewarde towarde the Center, or vpwarde from the Center, or circular about the Centre. Nowe vnto the earth and water in respect of their waight the motion downewarde is conuenient to seeke the Center. To ayre and fyer in regarde of their lightnes, vpwarde and from the Center. So is it meete to these elementes to attribute the right or streyghte motion, and to the heauens only it is proper circularly aboute this meane or Center to be turned rownde. Thus much *Aristotle*. Yf therefore saith *Ptolomy* of *Alexandria* the earth should turne but only by ye dayly motion, thinges quite contrary to these should happen. For his motion should be most swift and violent that in .24. howres should let passe the whole circuite of the earth, and those things whiche by sodaine toorninge are stirred, are altogether vnmeete to collecte, but rather to disperse thinges vnited, onelesse they shoulde by some firme fasteninge be kept toogether. And longe ere this the Earthe beinge dissolued in peeces should haue been scattered through ye heauens, which were a mockery to thincke of, and mutch more beastes and all other waights that are loose could not remayne vnshaken. And also things fallinge should not light on the places perpendiculare vnder theym, neyther shoulde they fall directly thereto, the same beinge violentlye in the meane carryed awaye. Cloudes also and other thynges hanginge in the ayre shoulde always seeme to vs to bee carried towarde the West.

The Solution of these Reasons
with their insufficiencye.

These are ye causes and sutch other wherwith they approue ye Earthe to reste in the middle of the worlde and that out of all question: But hee that will mainteyne the Earthes mobility may say that this motion is not violent but naturall. And these thinges whyche are naturally mooued haue effectes contrary to sutch as are violentlye carried. For sutche motions wherein force and vyolence is vsed, muste needes bee dissolued and cannot be of longe continuance, but those which by nature are caused, remayne stil in their perfit estate and are conserued and kepte in their moste excellent constitution. Without cause therefore did *Ptolomey* feare least the Earth and all earthelye thynges shoulde bee torne in peeces by thys reuolution of the Earthe caused by the woorkinge of nature, whose operations are farre different from those of Arte or sutche as humayne intelligence may reache vnto. But whye shoulde hee not mutch more thincke and misdought the same of the worlde, whose motion muste of necessity be so mutche more swift and vehemente then this of the Earth, as the Heauen is greater then ye Earth. Is therefore the Heauen made so huyge in quantitye that yt might wyth vnspeakeable vehemencye of motion bee seuered from the Centre, least happily restinge it should fall, as some Philosophers haue affirmed: Surelye yf this reason shoulde take place, the Magnitude of the Heauen shoulde infinitely extende. For the more this motion shoulde violentlye bee carryed higher, the greater should the swiftnes be, by reason of the increasing of the circumferēce which must of necessity in 24. houers bee paste ouer, and in lyke maner by increase of the motion the Magnitude muste also necessarilye bee augmented. Thus shoulde the swiftnes increase the Magnitude and the Magnitude the swiftnes infinitely: But according to that grounde of nature whatsoeuer is infinite canne neuer be passed ouer. The Heauen therefore of necessity must stande and rest fixed. But say they without the Heauen there is no body, no place, no emptynes, no not any thinge at all whether heauen should or could farther extende. But this surelye is verye straunge that nothinge shoulde haue sutche efficiente power to restrayne some thinge the same hauinge a very essence and beinge. Yet yf wee would thus confesse that the Heauen were indeede infinite vpwarde, and onely fynyte downewarde in respecte of his sphericall concauitye, Mutch more perhappes might that sayinge be verified, that without the Heauen is nothinge, seeinge euerye thinge in respect of the infinitenes thereof had place sufficient within ye same. But then must it of necessity remaine immoueable. For the cheefest reason that hath mooued some to thincke the Heauen limited was Motion, whiche they thoughte without controuersie to bee in deede in it. But whether the worlde haue his boundes or bee in deed infinite and without boundes, let vs leaue that to be discussed of Philosophers, sure we are y^t the Earthe is not infinite but hath a circumference lymitted, seinge therefore all Philosophers consent that lymitted bodyes maye haue Motion, and infinyte cannot haue anye. Whye dooe we yet stagger to confesse motion in the Earth beinge most agreeable to hys forme and nature, whose boundes also and circumference wee knowe, rather then to imagyne that the whole world should sway and turne, whose ende we know not, ne possibly can of any mortall man be knowne. And therefore the true Motion in deede to be in the Earth, and the apparâce only in the Heauen: And that these apparances are no otherwise then yf the *Virgilian Æneas* shoulde say.

Prouehimur portu, terræque vrbesque recedunt[4]

For a shippe carryed in a smoothe Sea with sutch tranquility dooth passe away, that al things on the shores and the Seas to the saylers seeme to mooue and themselues only quietly to rest with all sutche thinges as are aboorde them, so surely may it bee in the Earth whose Motion beinge naturall and not forcible of all other is most vniforme and vnperceaueable, whereby too vs that sayle therein the whole worlde maye seeme too roull about. But what shall wee then saye of Cloudes and other thinges hanginge or restinge in the ayre or tendinge vpward, but that not only the Earth and Sea makinge one globe but also no small part of the ayre is likewyse circularly carried and in like sort all sutche thinges as are deriued from them or haue any maner of aliance with them. Either for that the lower Region of the ayre beinge mixte with Earthlye and watrye vapours folowe the same nature of the Earth. Eyther that it be gayned and gotten from the Earth by reason of *Vicinity* or *Contiguity*. Whiche if any man merueyle at, let him consider howe the olde Philosophers did yeelde the same reason for the reuolution of the highest Region of the ayre, wherein we may sometime beholde Comets carryed circularly no otherwise then the bodies Celestial seeme to bee, and yet hath that Region of the ayre lesse conuenience with the Orbes Celestiall, then this lower part with the earthe, But we affyrme that parte of the aire in respect of his great distance to be destitut of this Motion *Terrestriall*, and that this part of the ayre that is next to ye Earthe dooth appeare moste still and quiet by reason of hys vniforme naturall accompanyinge of the

Earth, and lykewyse thinges that hange therein, onelesse by windes or other violent accident they be tossed to and fro. For the wynde in the ayre, is nothinge els but as the waue in the Sea: And of thinges ascēdinge and descendinge in respect of the worlde we must confesse them to haue a mixt motion of right & circulare, albeit it seeme to vs right & streight, No otherwise then if in a shippe vnder sayle a man should softly let a plūmet downe from the toppe alonge by the maste euen to the decke: This plummet passing alwayes by the streight maste, seemeth also too fall in a righte line, but beinge by discours of reason wayed his Motion is found mixt of right and circulare. For sutch thinges as naturally fall douneward beinge of earthly nature there is no doubt but as partes they retayne the nature of the whole. No otherwise is it of these things that by fiery force are carried vpward. For the earthly fyer is cheefly noorished wyth earthly matter, and flame is defined to be nought els but a burninge fume or smoke and the propertye of fyer is to extende the subject whereunto it entereth, the whiche it doth with so greate violence as by no meanes or engines it canne be constrayned but that with breache of bandes it will perfourme his nature. This motion extensiue is from the Centre to the circumference, so that if any earthly part be fiered, it is carryed violently vpward. Therefore whereas they say that of simple bodyes the motion is altogether simple, of the circulare it is cheefely verified, so longe as the simple bodye remayneth in his naturall place and perfit vnity of composition, for in the same place there can bee no other motion but circulare, whiche remayninge wholye in it selfe is most like to rest and immobility. But right or streight motion only happen to those thinges that stray and wander or by anye meanes are thrust out of their natural place. But nothing can bee more Repugnaunte to the fourme and Ordinance of the world, then that thinges, naturally should be out of their naturall place. This kinde of motion therefore that is by right line is only accident to those things that are not in their right state or perfection naturall, while partes are disioyned from their whole bodie, and couet to retourne to the vnity thereof againe. Neither do these thinges which are carryed vpwarde or downwarde besides this circular mouinge make anye simple, vniforme, or equall motion, for with their leuity or ponde-rositye of their body they cannot be tempered but alwaies as they fall (be-ginninge slowly) they increase their motion, and the farder ye more swiftly, whereas contrariwise this our earthly fier (for other wee cannot see) we may behould as it is carryed vpwarde to vanish and decay as it were confessinge the cause of violence to proceede only from his matter *Terrestriall*. The circulare motion alwaye contynueth vnyforme and equall by reason of his cause whiche is indeficient and alway continuinge. But the other hasteneth to ende and to attayne that place where they leaue lenger to be heuye or lighte, and hauinge attayned that place, theyr motion ceaseth. Seinge therefore this circulare motion is proper to the whole as streighte is only vnto partes, we may say that circulare doth rest with streighte as *Animall cum Aegro*. And whereas *Aristotle* hath dystry-buted *Simplicem motum* into these thre kyndes *A medio. ad medium*, and *Circa medium*, it must be onely in reason and imagination, as wee likewise seuer in consideration Geometricall a poincte, a line, and a superficies, whereas in deede neither can stand without other, ne any of them without a bodye.

Heereto wee may adioyne that the condition of immobilitye is more noble and diuine then y^t of chandge, alteration, or instabilitye, & therefore more agreeable to Heauen then to this Earth where al thinges are subiect to continual mutability. And seeinge by euident proofe of Geometricall mensuration wee finde that the Planets are sometimes nigher to vs and sometimes more remote, and that therefore euen the mainteyners of the Earthes stability are enforced to confesse that the Earth is not their Orbes Centre, this motion *Circa medium* must in more generall sort bee taken and that it maie bee vnderstande that euery Orbe hath his peculiare *Medium* and Centre, in regarde whereof this simple and vniforme motion is to bee considered. Seinge therefore that these Orbes haue seuerall Centres, it may be doughted whether the Centre of this earthly Grauity be also the Centre of the worlde. For Grauity is nothinge els but a certaine procliuitye or naturall coueting of partes to be coupled with the whole, whiche by diuine prouidence of the Creator of al is giuen & impressed into the parts, y^t they should restore themseules into their vnity and integritie concurringe in sphericall fourme, which kinde of propriety or affection it is likelye also that the Moone and other glorious bodyes wante not to knit & combine their partes together, and to mainteyne them in their round shape, which bodies notwithstandinge are by sundrye motions, sundrye wayes conueighed. Thus as it is apparant by these natural reasons y^t the mobility of the Earth is more probable and likelye then the stabilitye. So if it bee Mathematically considered and wyth Geometricall Mensurations euery part of euery *Theoricke* examined: the discreet Student shall fynde that *Copernicus* not without greate reason did propone this grounde of the Earthes Mobility.

Halley on the Infinity of the Sphere of Stars

FIRST PAPER

Of the Infinity of the Sphere of Fix'd Stars

Read before the Royal Society in March 1721, published in the *Philosophical Transactions* (1720–1721)[1]

The System of the World, as it is now understood, is taken to occupy the whole *Abyss* of *Space*, and to be as such actually infinite; and the appearance of the Sphere of Fixt Stars, still discovering smaller and smaller ones, as you apply better Telescopes, seems to confirm this Doctrine. And indeed, were the whole System finite; it, though never so extended, would still occupy no part of the *infinitum* of Space, which necessarily and evidently exists; whence the whole would be surrounded on all sides with an infinite *inane*, and the superficial stars would gravitate towards those near the center, and with an accelerated motion run into them, and in process of time coalesce and unite with them into one. And, supposing Time enough, this would be a necessary consequence. But if the whole be Infinite, all the parts of it would be nearly *in aequilibrio*, and consequently each fixt Star, being drawn by contrary Powers, would keep its place; or move, till such time, as, from such an *aequilibrium*, it found its resting place; on which account, some, perhaps, may think the Infinity of the Sphere of Fixt Stars no very precarious Postulate.[2]

But to this I find two Objections, which are rather of a Metaphysical than Physical Nature; and first, this supposes, as its consequent, that the number of Fixt Stars is not only indefinite, but actually more than any finite Number; which seems absurd *in terminis*, all Number being composed of Units, and no two Points or Centers being at a distance more than finite. But to this it may be answer'd, that by the same Argument we may conclude against the possibility of eternal Duration, because no number of Days, or Years, or Ages, can compleat it. Another Argument I have heard urged, that if the number of Fixt Stars were more than finite, the whole superficies of their apparent Sphere would be luminous, for that those shining Bodies would be more in number than there are Seconds of a Degree in the *area* of the whole Spherical Surface, which I think cannot be denied. But if we suppose all the Fixt Stars to be as far from one another, as the nearest of them is from the Sun; that is, if we may suppose the Sun to be one of them, at a greater distance their Disks and Light will be diminish'd in the proportion of Squares, and the Space to contain them will be increased in the same proportion; so that in each Spherical Surface the number of Stars it might contain, will be as the Biquadrate of their distances.[3] Put then the distances immensely great, as we are well assured they cannot but be, and from thence by an obvious *calculus*, it will be found, that as the Light of the Fix'd Stars diminishes, the intervals between them decrease in a less proportion, the one being as the Distances, and the other as the Squares thereof, reciprocally.[4] Add to this, that the more remote Stars, and those far short of the remotest, vanish even in the nicest Telescopes, by reason of their extream minuteness; so that, tho' it were true, that some such Stars are in such a place, yet their Beams, aided by any help yet known, are not sufficient to move our Sense; after the same manner as a small Telescopical fixt Star is by no means perceivable to the naked Eye.[5]

SECOND PAPER

Of the Number, Order, and Light of the Fix'd Stars

Read before the Royal Society in March 1721, published in the *Philosophical Transactions* (1720–1721)

At the last meeting of the Society, I adventured to propose some Arguments, that seemed to me to evince the Infinity of the Sphere of Fixt Stars, as occupying the whole Abyss of Space, or the τὸ πᾶν, which at present is generally understood to be necessarily Infinite; and thence I laid before you what may seem a very *Metaphysical Paradox, viz.* That the number of Fixt Stars must then be more than any finite Number, and some of them more than at a finite distance from others. This seems to involve a Contradiction, but it is not the only one that occurs to those who have undertaken freely to consider the nature of Infinite, which perhaps the very narrow limits of humane Capacity cannot attain to.

Since then, I have attentively examined what might be the consequence of an Hypothesis, that the Sun being one of the Fixt Stars, all the rest were as far distant from one another, as they are from us; and by a due calculation I find, that there cannot, upon that Supposition, be more than 13 Points in the Surface of a Sphere, as far distant from the Center of it, as they are from one another; and I believe it would be hard to find how to place thirteen Globes of equal magnitude, so as to touch one another in the Center: for the twelve Angles of the *Icosaedron* are from one another very little more distant than from its center; that is, the side of the Triangular Base of that Solid, is very little more than the Semidiameter of the circumscribed Sphere, it being to it nearly as 21 to 20; so that it is plain that somewhat more than twelve equal Spheres may be posited about a middle one; but the Spherical Angles or Inclinations of the planes of these Figures being incommensurable with the 360 degrees of the Circle, there will be several interstices left, between some of the Twelve, but not such as to receive in any part the thirteenth Sphere.[6]

Hence it is no very improbable Conjecture, that the number of the Fixt Stars of the first magnitude is so small, because this superior appearance of Light arises from their nearness; those that are less shewing themselves so small by reason of their greater distance. Now there are in all but sixteen Fixt Stars, in the whole number of them, that can indisputably be accounted of the first magnitude; whereof four are *extra Zodiacum;* viz. *Capella, Arcturus, Lucida Lyrae,* and *Lucida Aquilae,* to the *North;* four in the way of the *Moon* and *Planets,* to wit, *Palilicium, Cor Leonis, Spica,* and *Cor Scorpii;* and five to the *Southward,* that are seen in England, viz. The *Foot* and Right *Shoulder* of *Orion, Sirius, Procyon,* and *Formalhaut;* and there are three more that never rise in our Horizon, viz. *Canopus, Acharnâr,* and the *Foot* of the *Centaur.* But that they exceed the number Thirteen, may easily be accounted for from the different magnitudes that may be in the Stars themselves; and perhaps some of them may be much nearer to one another, than they are to us; this excess of Number being found singly in the Signs of *Gemini* and *Cancer.* And indeed within 45 degrees of Longitude, or one 8th of the whole, there are no less than *five* of these *sixteen* to be seen. If therefore the Number of them be supposed *Thirteen,* omitting Niceties in a Matter of such Irregularity, at twice the distance from the Sun there may be placed four times as many, or 52; which, with the same allowance, would nearly represent the number of the Stars we find to be of the 2d magnitude: so 9×13, or 117, for those at three times the distance: and at ten times the distance 100×13 or 1300 Stars; which distance may perhaps diminish the light of any of the Stars of the first magnitude to that of the sixth, it being but the hundredth part of what, at their present distance, they appear with.[7] But if, since we have room enough for it, we should suppose the Sphere continued to 10 times the last, or 100 times the first distance, the number of Stars would be 130 000, and they would appear but with the 10 000th part of the Light of a first magnitude Star, as we now see it. This is so small a pulse of Light, that it may well be questioned, whether the Eye, assisted with any artificial help, can be made sensible thereof. But 100 times the distance of a Star we see, is still Finite: from whence I leave those that please to consider it attentively, to draw the Conclusion.[8]

Chéseaux Explains the Riddle
of Darkness

On the force of light and its propagation in the ether, and the distances to the fixed stars

Traité de la Comète qui a Paru en Décembre 1743 & en Janvier, Février & Mars 1744
(Lausanne: Bousquet et Compagnie, 1744), appendix 2, pp. 223–229[1]

If all the fixed stars were so many Suns, similar and equal to our own, when placed at the same distance they would have the same apparent size and luminosity as the Sun, and would send the same amount of light to us. According to a proposition in optics the amount of light sent to us by a star, no matter how far from the Earth, equals the direct ratio of the square of its apparent diameter to the square of that of the Sun, or the inverse ratio of the square of its distance to the square of that of the Sun.[2] Let us now imagine that all of starry space is divided into concentric spherical layers of approximately constant thickness equal to the diameter of the vortex[3] of the planetary system of each star; also let the number of stars in a layer be approximately proportional to the surface area of that layer, or to the square of its distance from the Sun, with the Sun taken as the center of the starry heavens; finally, let the actual diameter of each star be approximately equal to that of the Sun. As I have said, we will find that the amount of light that is sent to us by the stars in any single layer is proportional to the sum of the squares of their apparent diameters, that is, proportional to the number of stars in each layer multiplied by the square of the apparent diameter of any of these stars, or in accordance with what I have just said, proportional to the square of the distance of a layer divided by this same square;[4] and thus the amount of light is the same from all layers. The amount of light from each layer equals the amount we receive from the Sun times the ratio of the square of the distance of the Sun from the Earth to the square of the distance of the first layer of stars divided by the number of stars in that layer,[5] that is, approximately the ratio of 1 to 4,000,000,000. From this argument it follows that if starry space is infinite, or only larger than the volume occupied by the Solar System and the first-magnitude stars by the ratio of the cube of 760,000,000,000,000 to 1, each bit of the sky would appear as bright to us as any bit of the Sun, and therefore the amount of light received from each celestial hemisphere—one above and the other below the horizon—would be 91,850 times greater than what we receive from the Sun. The enormous difference between this conclusion and experience demonstrates either that the sphere of fixed stars is not infinite but actually incomparably smaller than the finite extension I have supposed, or that the force of light decreases faster than the inverse square of distance. This latter supposition is quite plausible; it requires only that starry space is filled with a fluid capable of intercepting light very slightly. Even if this fluid were 330,000,000,000,000,000 times more transparent or thinner than water, this would suffice to weaken the force of light by one thirty-third as it passes through each layer. All the light sent to us beyond the layers closest to our vortical system would be gradually absorbed to the point where the total amount of light from one hemisphere is reduced to 1 part in 430,000,000 the amount of light from the Sun, or to an amount only 33 times greater than the light received from the dim globe of the new Moon lighted by the Earth.[6]

No doubt you will think that the numbers I have used were chosen by guesswork. This is true, but my conjectures were not entirely arbitrary, at least not the distance to the fixed stars of first magnitude. I have estimated this distance to be 240,000 times greater than the distance of the Sun according to certain principles. On May 16, 17, and 18, 1743, at the time of the conjunction of Mars with Saturn, I noticed that their brilliance exceeded that of any fixed star of first magnitude, even of Sirius. By considering their distances from the Earth and the Sun, their apparent size and dichotomous shape, and the brightness of their light compared with that of the Moon and Sun, I determined that the apparent diameter of a star of first magnitude is less than 1/119 of a second of arc. During the conjunction of Mars with Jupiter on June 1 and 2, the brilliance of Mars seemed equal to that of Regulus, or a little less than that of the most brilliant stars. Using the same procedure as before, I found that the apparent diameter of the fixed stars of first magnitude is slightly more than 1/131 of a second of arc. On the average the diameter of these stars is 1/125 of a second of arc, or 7 0° 0' 0" 0''' 28^{iv} 48^{v}. I compared this value with the diameter of the Sun, used 15" for the parallax of the Sun, and employed the method used by the learned academician Mr. Bouguer for determining the opacity of water in his essay on the gradation of light.[8] This procedure gave me, with hardly any other help, all my results, except for the hypothesis I was obliged to make concerning the amount of light sent by the stars in the first layer compared with the amount sent by the entire firmament. The diameter of Jupiter at perigee is 50", also its distance from the Sun is 5 times the Moon's distance from the Sun; consequently the intensity of its light is 7,500,000 times weaker than that of the Sun or fixed stars. From the apparent diameter of 0° 0' 0" 0''' 28^{iv} 48^{v} of the first magnitude stars, the brilliance of Jupiter (in these circumstances) would equal that of 5 such stars, or 2/5 or 1/3 of all first magnitude stars taken together. Now the brilliance of that planet seemed to me—as well as I could judge by rough experiments comparing certain objects illumined by a section of the firmament with those same objects illumined by the planet—to be approximately one 50th that of an entire hemisphere, or one 100th that of the whole firmament. From this I concluded that the latter—the light of the entire firmament—to be 33 times greater than that from stars of first magnitude in the first layer. The brilliance of Venus at half phase, at its mean distance, seems four times greater than the brilliance of Jupiter, and therefore, according to this hypothesis, is one twelfth the brilliance of an entire hemisphere.

Olbers Revives the Riddle
of Darkness

On the transparency of space

Submitted 7 March 1823 by Dr. Olbers of Bremen, *Astronomisches Jahrbuch für das Jahr 1826* (Berlin: C. F. E. Späthen, 1823), pp. 110–121[1]

Large and small in space are of course relative only. We can imagine creatures to whom a grain of sand would seem as large as the Earth seems to us; also we can conceive of another order in which bodies surpassing planets and suns in size would seem no larger than grains of sand. For this reason it is natural for man to judge whether things are great or small according to the scale of his own body and the bodies around him. In this way man evaluates the magnitude of things and consequently views with astonishment the vastness of the universe revealed by telescopes. The distance of the Sun from the Earth is already so great that attempts have been made to render it more comprehensible by calculating the time taken for a cannon ball to travel to the Sun. Every fixed star is a sun. But the nearest star is so far away that by comparison the distance of the Sun from the Earth dwindles to almost nothing. With the naked eye we observe numerous stars of various magnitudes ranging from bright Sirius to those of sixth and seventh magnitude that can only just be detected by the sharpest eye on the clearest night. Some of these faint stars may actually be quite small; most, however, appear faint because of their great distance, and with the naked eye we perceive stars that are twelve to fifteen times farther away than those of first magnitude. Improvements in telescopes continually reveal fainter stars, even though we find it difficult to comprehend the vast distances and reaches of space revealed by Herschel in his giant telescope when he explored the heavens out to distances 1500 and even several thousand times more remote than Sirius or Arcturus.

But did the late keen-eyed Herschel penetrate to the outer limits of the universe? Did he even succeed in approaching these limits? Who can believe that he did? Is space not infinite? Are limits to it conceivable? And is it conceivable that the omnipotence of the Creator would have left this interminable space empty? I will let the celebrated Kant speak for me:[2] "Where shall creation itself cease?" says Kant. "It is evident that in order to think of it as in proportion to the power of the Almighty, it must have no limits at all. We come no nearer to the infinitude of God's creative power by enclosing space within a sphere having the radius of the Milky Way than by limiting space to a small ball an inch diameter. All finite things possessing limits and definite relations to unity are equally far removed from infinity. It would be absurd to represent the Deity as manifesting finite creative power; it would be illogical to think of divine infinite power—that limitless storehouse of natures and worlds—existing in a permanently inactive state. Surely it seems reasonable and even necessary to regard the created universe as evidence of that power that cannot be measured by any standard whatever? For this reason the field of divine nature reveals itself as infinite as that nature itself. Eternity alone fails to embrace the manifestations of the supreme being and must be conjoined with an infinitude of space."

So reasoned Kant. Quite probably not only the portion of space that our eye has penetrated with the aid of instruments, or may be penetrated in the future, but also endless space itself is sprinkled with suns and their accompanying planets and comets. I say "probably," but our limited reason guarantees no certainty. Indeed other parts of space may contain creations entirely different from suns, planets, and comets, having an entirely different form of light; creations perhaps altogether beyond our imagination. Halley has attempted to establish the existence of an infinity of stars. "If their number were not infinite," he says, "there would exist a center of gravity[3] at a certain point within the space occupied by the stars. All bodies in space would fly toward this point with increasing speed and the universe would thus collapse upon itself. Only because the universe is infinite can the whole maintain itself in equilibrium." Halley considered gravitational forces only and neglected centrifugal forces. If there were no fixed stars, and only our planetary system existed, the planets would not fall into the Sun. The motions of the stars actually indicate that centrifugal forces cannot be ignored. This fact alone suffices to show the inadequacy of Halley's argument; other errors also refute his argument.

Nevertheless, in spite of the inadequacies of Halley's treatment, it remains highly probable that this sublime order, extending as far as the eye can reach, continues much the same throughout infinite space. Let us try to discover arguments that might compel us to deny this hypothesis. Immediately an important objection springs to mind. If there really are suns throughout the whole of infinite space, and if they are placed at equal distances from one another, or grouped into systems like that of the Milky Way, their number must be infinite and the whole vault of heaven must appear as bright as the Sun; for every line that we can imagine drawn from our eyes would necessarily lead to some fixed star, and therefore starlight, which is the same as sunlight, would reach us from every point of the sky.

It goes without saying that experience contradicts this argument. Halley rejects the argument that with an infinite number of stars the sky must blaze as bright as the Sun at every point. But he does so on altogether erroneous grounds. He confuses the apparent and absolute magnitudes of stars, and arrives at the conclusion that the number of fixed stars increases as the square of their distance from us, whereas the space between them increases as the fourth power of distance. This conclusion is false. Let us assume that the fixed stars are uniformly distributed throughout space. We imagine a sphere, with the Sun at the center, having a radius equal to the average distance of the stars of first magnitude. Let this radius have unit value. Furthermore, let N_1 denote the number of first magnitude stars, and let δ be the radius of a star. We find that the nearest stars cover $\frac{1}{4}N_1\delta^2$ of the celestial vault.[4] The fixed stars at distance 2, of number $4N_1$, have an apparent diameter $\frac{1}{2}\delta$, and also cover $\frac{1}{4}N_1\delta^2$ of the celestial vault. Hence, stars at distances 1, 2, 3, 4, 5, . . . , m cover equal areas of the celestial vault, and their sum

$$\frac{1}{4}N_1\delta^2 + \frac{1}{4}N_1\delta^2 + \frac{1}{4}N_1\delta^2 \ldots = \frac{1}{4}mN_1\delta^2$$

becomes infinitely great as m goes to infinity, regardless of how small is the quantity $\frac{1}{4}\delta^2$. Not only will the whole celestial vault be covered with stars, but also stars will stand one behind another in endless rows covering each other from our view. Clearly, whether stars are uniformly distributed in space or clustered in widely separated systems, the same conclusion follows.

How fortunate for us that nature has arranged matters differently! How fortunate that the Earth does not receive starlight from every point of the celestial vault! Yet, with such unimaginable brightness and heat, amounting to 90,000 times more than what we now experience, the Almighty could easily have designed organisms capable of adapting to such extreme conditions. But astronomy for the inhabitants of the Earth would remain forever in a primitive state; nothing would be known about the fixed stars; only with difficulty would the Sun be detected by virtue of its spots; and the Moon and planets would be distinguished merely as darker disks against a background as brilliant as the Sun's disk.

Because the celestial vault has not at all points the brightness of the Sun, must we reject the infinity of the stellar system? Must we restrict this system of stars to one small portion of limitless space? Not at all. In our inferences drawn from the hypothesis that an infinite number of fixed stars exists, we have assumed that space throughout the whole universe is absolutely transparent, and that light, consisting of parallel rays, remains unimpaired as it propagates great distances from luminous bodies. This absolute transparency of space, however, is not only undemonstrated but also highly improbable. Because planets, which have high density, do not encounter any noticeable resistance in their motion, we should not assume that space is entirely empty. What we know of comets and their tails suggests the presence of something material in the regions they traverse. Also the matter composing comet tails, which gradually disperses, and also the matter responsible for zodiacal light provide further evidence. Even if space were quite empty, the crossing of light rays would cause small losses.[5] This can be proved not only by a priori arguments, using either Newton's or Huygen's theory of light, but also by experiments contrasting the performance of the Cassegrain and Gregorian telescopes and comparing the density of light in front of and behind the focus of spherical mirrors.*

*Philosophical Transactions (1813–1814). Captain Kater in calculations of the relative density of light in front of and behind the focus of concave mirrors appears to have overlooked the fact that a so-called focus must be regarded not as a physical point but only as an image of the Sun or the candle flame. The corrections necessary, however, need not invalidate his conclusion that light suffers a small loss while passing through a focus. It is desirable that these interesting experiments be repeated with greater care.

Without doubt the universe is not absolutely transparent. Only the slightest degree of non-transparency suffices to refute the conclusion—so contrary to our experience—that if the fixed stars stretch away to unlimited distance, the entire sky must blaze with light. Assume, for example, that space is transparent to the extent that of every 800 rays emitted by Sirius, 799 survive after traveling a distance equal to that separating us from Sirius. This small amount of absorption in an endless star-filled universe is more than sufficient to create the conditions observed on Earth . . .[6]

We may safely say, given the assumed amount of absorption, that all stars more than 30,000 times the distance of Sirius contribute nothing to the brightness of the sky.[7]

But for the fact that our atmosphere has a certain brightness, even when illuminated solely by stars, the heavens would appear completely black. Because of atmospheric light the sky appears dark blue and never completely black even on the clearest nights . . .[8]

The assumption that the light coming to us from Sirius has been weakened by one part in 800, independently of its divergence, is of course quite arbitrary. As I mentioned, I intended to show by the use of this value that a small loss of light, or an even smaller loss, would suffice to explain the observed darkness of the night sky, despite the possibility that the number of stars is infinite in endless space. The amount of non-transparency was not chosen entirely at random, however, and I believe that it corresponds quite closely to the actual amount prevailing.

The Almighty with benevolent wisdom has created a universe of great yet not quite perfect transparency, and has thereby restricted the range of vision to a limited part of infinite space. Thus we are permitted to discover some of the design and construction of the universe, of which we could know almost nothing if the remotest suns were allowed to blaze with undiminished light.

Kelvin on an Old and Celebrated Hypothesis

On ether and gravitational matter
through infinite space

First published in *Philosophical Magazine*, series 6, vol. 2, pp. 161–177 (1901);
reprinted in *Baltimore Lectures on Molecular Dynamics and the Wave Theory of
Light* (London: Cambridge University Press, Clay & Sons, 1904), Lecture 16,
pp. 260–278

18. Many of our supposed thousand million stars, perhaps a great majority of
them, may be dark bodies; but let us suppose for a moment each of them to be
bright, and of the same size and brightness as our sun; and on this supposition
and on the further suppositions that they are uniformly scattered through a
sphere of radius 3.09 10^{16} kilometers, and that there are no stars outside this
sphere, let us find what the total amount of starlight would be in comparison
with sunlight. Let n be the number per unit of volume, of an assemblage of
globes of radius a scattered uniformly through a vast space. The number in a
shell of radius q and thickness dq will be $n.4\pi q^2 dq$, and the sum of their apparent areas as seen from the centre will be[1]

$$\frac{\pi a^2}{q^2} n.4\pi q^2 dq \quad \text{or} \quad n.4\pi^2 a^2 dq.$$

Hence by integrating from $q = 0$ to $q = r$ we find

$$n.4\pi^2 a^2 r \qquad (8)$$

for the sum of their apparent areas. Now if N be the total number in the sphere
of radius r we have

$$n = N/\left(\frac{4\pi}{3}r^3\right) \qquad (9)$$

Hence (8) becomes $N.3\pi(a/r)^2$; and if we denote by α the ratio of the sum of the
apparent areas of all the globes to 4π we have

$$\alpha = \frac{3N}{4}\left(\frac{a}{r}\right)^2 \qquad (10)$$

$(1 - \alpha)/\alpha$, very approximately equal to $1/\alpha$, is the ratio of the apparent area not
occupied by stars to the sum of the apparent areas of all their discs. Hence α is
the ratio of the apparent brightness of our star-lit sky to the brightness of our

sun's disc.[2] Cases of two stars eclipsing one another wholly or partially would, with our supposed values of r and a, be so extremely rare that they would cause a merely negligible deduction from the total of (10), even if calculated according to pure geometrical optics. This negligible deduction would be almost wholly annulled by diffraction, which makes the total light from two stars of which one is eclipsed by the other, very nearly the same as if the distant one were seen clear of the other.[3]

19. According to our supposition of §§18 we have $N = 10^9$, $a = 7 \ 10^5$ kilometers, and therefore $r/a = 4.4 \ 10^{10}$. Hence by (10)

$$\alpha = 3.87 \ 10^{-13} \tag{11}$$

This exceedingly small ratio will help us to test an old and celebrated hypothesis that if we could see far enough into space the whole sky would be seen occupied with discs of stars all of perhaps the same brightness as our sun, and that the reason why the whole of the night-sky is not as bright as the sun's disc is that light suffers absorption in traveling through space. Remark that if we vary r keeping the density of the matter the same, N varies as the cube of r. Hence by (10) α varies simply as r; and therefore to make α even as great as 3.87/100, or, say, the sum of the apparent areas of discs 4 per cent. of the whole sky, the radius must be $10^{11}r$ or $3.09 \ 10^{27}$ kilometers. Now light travels at the rate of 300,000 kilometers per second or $9.45 \ 10^{12}$ kilometers per year. Hence it would take $3.27 \ 10^{14}$ or about $3\frac{1}{4} \ 10^{14}$ years to travel from the outlying suns of our great sphere to the centre. Now we have irrefragable dynamics proving that the whole life of our sun as a luminary is a very moderate number of million years, probably less than 50 million, possibly between 50 and 100. To be very liberal, let us give each of our stars a life of a hundred million years as a luminary.[4] Thus the time taken by light to travel from the outlying stars of our sphere to the centre would be about three and a quarter million times the life of a star. Hence, if all the stars through our vast sphere commenced shining at the same time, three and quarter million times the life of a star would pass before the commencement of light reaching the earth from the outlying stars, and at no one instant would light be reaching the earth from more than an excessively small proportion of all the stars. To make the whole sky aglow with the light of all the stars at the same time the commencements of the different stars must be timed earlier and earlier for the more and more distant ones, so that the time of arrival of every one of them at the earth may fall within the durations of the lights at the earth of all the others! Our supposition of uniform density of distribution is, of course, quite arbitrary; and we ought in the greater sphere to assume the density much smaller than in the smaller sphere; and in fact it seems there is no possibility of having enough of stars (bright or dark) to make a total of star-disc-area more than 10^{-12} or 10^{-11} of the whole sky.

★

Notes

Full publication information for the works cited in the notes
is given in the Bibliography.

1. Why Is the Sky Dark at Night?

1. The sky has an angular area of 4π square radians (a square radian is a unit of solid angle, the steradian). A radian equals $180/\pi = 57.3$ arc degrees, and the whole sky is covered with $4 \times 180^2/\pi = 41,253$ square degrees. The Sun subtends an angular radius of almost 0.27 degrees, corresponding to an area little more than 0.22 square degrees. The area of the whole sky is therefore roughly 180,000 times that of the Sun. In other words, a bright-sky universe pours down on Earth 180,000 times as much radiation as the Sun.

2. See Wesley Salmon, *Zeno's Paradoxes*. Essays by several authors on Zeno's paradoxes may be found in Richard Gale, *The Philosophy of Time*, chapter 5. The paradoxes are discussed in G. Whitrow, *The Natural Philosophy of Time*, pp. 190–200, and also in Bertrand Russell, "Mathematics and metaphysicians," in *Mysticism and Logic*.

3. See Silvanus P. Thompson, *The Life of William Thomson, Baron Kelvin of Largs* vol. 2, p. 833. Stanley Jaki, *The Paradox of Olbers' Paradox*, traces the history of the dark night-sky riddle back to Halley. Jaki argues that the paradox of the dark night sky is itself a many-stranded paradox consisting of scientists failing to realize the importance of the paradox, failing to consult the historical records, and failing to perceive the correct solution. While on this theme, one might also mention the paradox of historians of natural science knowing little or no natural science. According to Jaki, the correct solution is as follows: Since the sky is covered with stars in an infinite universe, obviously it is uncovered in a finite universe; hence the universe must be finite. This argument, however, is false, as will be shown. The reasons why we must be cautious when calling the riddle a paradox are discussed in my "Dark night-sky riddle: a 'paradox' that resisted solution" (1984).

2. Three Rival Systems

1. *Plutarch's Essays and Miscellanies*, vol. 1, p. 140.
2. See Norman DeWitt, *Epicurus and His Philosophy*.
3. See the Introduction to *Lucretius on the Nature of Things*, in which Cyril

Bailey describes the Epicurean philosophy. The standard translation is Bailey's *Titi Lucreti Cari: De Rerum Natura.* In the Prolegomena of this work, Bailey gives the history and character of the manuscripts, all of which are different copies of an archytype in the fourth or fifth century. The most influential copies in the fifteenth and sixteenth centuries were the *Itali* deriving from the Poggio manuscript. The Poggio manuscript was published in Paris in 1564, and this publication served as the authoritative Lambinus edition for almost three centuries. In the text I quote from Ronald Latham's prose translation, *Nature of the Universe,* because of its easy style and ready availability.

4. See Samuel Sambursky, *The Physical World of the Greeks.*

5. See Samuel Sambursky, *Physics of the Stoics,* and David Hahm, *The Origins of Stoic Cosmology.*

6. Little information is available on Archytas. For a few comments see Max Jammer, *Concepts of Space;* B. Van der Waerden, *Science Awakening;* and Edward Grant, *Much Ado about Nothing.*

7. Quoted by Francis Cornford in "The invention of space."

8. Translated by Ronald Latham in *The Nature of the Universe,* pp. 55–56.

9. Translated by Edward Grant in "Medieval and seventeenth-century conceptions of an infinite void space beyond the cosmos."

10. Translated by Dorothea Singer in *Giordano Bruno: His Life and Thought,* p. 251.

11. John Locke, *An Essay Concerning Human Understanding,* bk. 2, chap. 13, sect. 21.

12. Dionys Burger, *Sphereland: A Fantasy about Curved Space and an Expanding Universe.* This book forms a sequel to the classic work by Edwin Abbott, *Flatland: A Romance of Many Dimensions.*

3. Celestial Light

1. See Lynn White, *Medieval Technology and Social Change,* and Jean Gimpel, *The Medieval Machine.*

2. Charles Haskins, *The Rise of Universities.*

3. Pierre Duhem (1861–1916), one of France's greatest intellects, was ignored by Parisian academics during most of his lifetime (see Donald Miller, "Ignored intellect: Pierre Duhem"). Duhem emphasized the great importance of medieval physics and cosmology, which previously had attracted very little attention from historians (see *Medieval Cosmology: Theories of Infinity, Place, Time, Void, and the Plurality of Worlds*). In "Late medieval thought, Copernicus, and the scientific revolution," Edward Grant quotes Duhem: "If we must assign a date for the birth of modern science, we would, without doubt, choose the year 1277 when the bishop of Paris solemnly proclaimed that several worlds could exist, and that the whole of the heavens could be moved with a rectilinear motion."

4. See Arthur Lovejoy, *Great Chain of Being,* chaps. 3 and 4.

5. Edward Grant translates and discusses various passages by Bradwardine

in "Medieval and seventeenth-century conceptions of an infinite void space beyond the cosmos."

6. Nicholas of Cusa, *On Learned Ignorance.*

7. Thomas Kuhn, *Copernican Revolution,* chap. 7.

8. Thomas Digges's work is treated in Francis Johnson, *Astronomical Thought in Renaissance England,* chap. 6; this chapter is reproduced in *Theories of the Universe,* edited by Milton Munitz, pp. 184–189.

9. Aided by Francis Johnson's publications, I have tentatively traced the riddle of darkness back to Thomas Digges in "The dark night-sky riddle: a 'paradox' that resisted solution." Digges participated in the rebirth of the idea of an infinite universe and raised the question concerning the visibility of distant stars. Ideas rarely have clear beginnings, and in Digges's work we see ideas forming that later, after the discovery of the telescope, become more clearly defined.

10. Dorothea Singer, *Giordano Bruno: His Life and Thought,* p. 302.

11. William Gilbert, *On the Magnet,* translated by Silvanas P. Thompson, pp. 215–216.

12. Marjorie Nicolson, *Science and Imagination,* p. 96, attributes this remark to David Masson in his lectures at Edinburgh. See her *Breaking of the Circle: Studies in the Effect of the "New Science" upon Seventeenth Century Poetry;* see also Thomas Orchard, *Milton's Astronomy: The Astronomy of 'Paradise Lost.'* Caroline Spurgeon, *Shakespeare's Imagery, and What it Tells Us,* p. 13, writes: "With Shakespeare, nature (especially the weather, plants and gardening), animals (especially birds), and what we may call everyday and domestic, the body in health and sickness, indoor life, fire, light, food and cooking, easily come first; whereas with Marlowe, images drawn from books, especially the classics, and from the sun, moon, planets and heavens far outnumber all others. Indeed this imaginative preoccupation with the dazzling heights and vast spaces of the universe is, together with a magnificent surging upward thrust and aspiration, the dominating note of Marlowe's mind." Nicholson reminds us that Marlowe, though under a new spell, is still an Elizabethan, whereas Donne has broken loose from the old order that now lies in ruins, and is the first English poet (for Marlowe died young) to grasp the implications of the new celestial discoveries.

4. The Starry Message

1. The blue sky of heaven in the medieval system is discussed in C. S. Lewis, *The Discarded Image.* Max Jammer, "Judeo-Christian ideas about space" (in *Concepts of Space*), discusses the belief that space is coextensive with light. Jammer writes that the belief originated with the Pythagoreans, and quotes from Plato's *Republic* (X, 616): "For this light binds the sky, like the hawser that strengthens the trireme, and thus holds together the whole revolving universe." The "metaphysics of light" woven into the mystery religions by Neoplatonists and Gnostics exerted a strong influence on scholars in the Middle Ages, including Robert Grosseteste, bishop of Lincoln, who advocated the theory that light is the primary element of the universe. See Alistair Crombie, *Robert Grosseteste and the*

Origins of Experimental Science, and Katherine Collier, "Primeval light," in *Cosmogonies of Our Fathers,* chap. 3.

2. See Edward Rosen, *The Naming of the Telescope.*

3. In the title *Sidereus nuncius* (*Starry Message* or *Sidereal Message,* the translation of "nuncius" as "messenger" gives the wrong sense. See E. Rosen, *Kepler's Conversation with Galileo's Sidereal Messenger,* pp. xiv–xvi; also "The title of Galileo's 'Sidereus nuncius.'" Stillman Drake translates Galileo's book under the title *The Starry Messenger* in his *Discoveries and Opinions of Galileo.* In the text I use the title *Starry Message,* and the quotations are from pp. 45–46, 47, and 49 of Drake's translation.

4. The astronomer John Herschel in 1836 measured a hundred-fold difference in brightness between first and sixth magnitudes. Norman Pogson proposed in 1850 that a unit decrease in magnitude corresponds to an increase in brightness by a factor 2.5 (more accurately, 2.512). Thus, $2.512^5 = 100$, and an increase in five magnitudes corresponds to a hundred-fold decrease in brightness. As a rough and ready guide, a candle at one kilometer distance is a first magnitude source of light, and at ten kilometers, when a hundred times fainter, a sixth magnitude source. Aldebaran and Altair are stars of the first magnitude, and the pole star Polaris is second magnitude. Sirius (magnitude -1.6) is 10 times brighter, Venus (magnitude -4) is 100 times brighter, and the full Moon (magnitude -12.5) is 250,000 times brighter than a star of first magnitude. A telescope of aperture one inch has a light-collecting area 16 times larger than the area of the pupil of a dark-adapted eye, and with this telescope we see stars of the ninth magnitude. This was about the aperture of Galileo's first telescope. A 100-inch telescope collects $100^2 = 10,000$ times more light and enables us to see by eye nineteenth-magnitude stars. Photographic plates increase the light-collecting power and further extend the magnitude range.

5. Galileo, *Dialogue Concerning the Two Chief World Systems—Ptolemaic and Copernican,* translated by Stillman Drake.

6. Alexandre Koyré, *From the Closed World to the Infinite Universe,* pp. 61, 69–70, 75, and 78.

7. The minimum angle resolvable by the eye is determined by diffraction due to the pupil and the separation of cone cells in the central retina. Consider a beam of rays entering the eye that would come to a perfect focus on the retina in the absence of diffraction. If λ represents the wavelength, d the diameter of the aperture (in this case the pupil), the diffraction pattern on the retina forms a central disk containing 0.84 of the incident light and subtending an angle 2.44 λ/d radians. (One radian equals $180/\pi = 57.3$ degrees, or 3,440 minutes.) With $\lambda = 4 \times 10^{-5}$ centimeters for blue light, and $d = 0.4$ centimeters, we obtain an angle 0.8 minutes. This corresponds roughly to the cone-cell separation in the central retina. A star appears as a tiny diffraction disk, and the size of this disk is very much larger than the actual geometric size of the star. Optical imperfections further enlarge the apparent size, and also atmospheric fluctuations cause scintillation (or twinkling). Planets and galaxies and other luminous objects that do not twinkle have angular sizes not greatly smaller than 1 arc minute. Tele-

scopes have larger apertures and can resolve smaller angles than the unaided eye. Objects such as stars subtending angles much smaller than the diffraction-limited angle cannot normally be magnified by telescopes. Telescopes serve as light-collectors and make a star look brighter but not larger. Kepler mistakenly thought that invisible stars become visible in the telescope because of magnification.

8. Kepler, *Conversation with the Starry Messenger,* translated by Edward Rosen; quotations from pp. 34, 35–36, and 43.

9. Kepler, *Epitome of Copernican Astronomy,* quoted by Koyré, *From the Closed World,* p. 78.

10. Alexandre Koyré, *From the Closed World,* pp. 86–87.

5. The Cartesian System

1. *A Discourse on the Method of Rightly Conducting the Reason and Seeking for Truth in the Sciences* (1637) and *Principles of Philosophy* (1644) are translated by Elizabeth Haldane and G. R. T. Ross in *The Philosophical Works of Descartes.* Quotations are from *Discourse on the Method,* pt. V, and *Principles,* pt. II, principle XX.

2. A universe of indefinitie extent governed by rational principles inspired Henry More to write in verse in *Infinite of Worlds* (1646), "Though I detest the sect of Epicurus for their manners vile, yet what is true I may not well reject." Margorie Nicolson, in "The early stages of Cartesianism in England," writes, "In the 1640s and 1650s the liberal theologians and philosophers saw in Descartes a savior;" but "by 1675 his philosophy was more often condemned than praised."

3. The work of Otto von Guericke is discussed by Edward Grant in "Medieval and seventeenth century conceptions of an infinite void space beyond the cosmos," also in *Much Ado about Nothing,* pp. 215–216. Grant translates parts of books I and II of *New Magdeburg Experiments on Void Space* in *A Source Book in Medieval Science,* pp. 563–568.

4. Neither Guericke nor Halley was the first to predict the periodicity of comets. Samuel Pepys wrote in his diary, March 1, 1665: "At noone I to dinner at Trinity House, and thence to Gresham College, where Mr. Hooke read a second very curious lecture about the late Comet. Among other things proving very probably that this is the same Comet that appeared before in the year 1618, and that in such a time probably it will appear again." Parts of books 1 and 2 of *New Magdeburg Experiments on Void Space* are translated from Latin by Edward Grant in *A Source Book in Medieval Science,* pp. 563–568. See also Grant's "Medieval and seventeenth-century conceptions of an infinite void space beyond the cosmos."

5. Descartes, *Dioptrique,* published in 1637. See A. I. Sabra, *Theories of Light from Descartes to Newton,* chaps. 1–4. See Newton, *Opticks,* edited by I. Bernard Cohen and Robert Schofield; and "The optical lectures, 1670–1672," edited by A. E. Shapiro.

6. The primary source of information on Robert Hooke (1635–1703) is

Richard Waller's brief account prefixed to *The Posthumous Works of Robert Hooke*, published in 1705. "Robert Hooke" by Edward Andrade summarizes much of what is now known about Hooke. The remarkable *Micrographia: Or Some Physiological Descriptions of Minute Bodies Made by Magnifying Glasses with Observations and Inquiries Thereupon*, 1665, by Hooke at age thirty, with numerous illustrations, contains many original thoughts, such as, "Heat is a property of a body arising from the motion or agitation of its parts," and light "a very short vibrative motion transverse to straight lines of propagation through a homogeneous medium." Marjorie Nicolson describes in her essays in *Science and the Imagination* and lectures in *The Breaking of the Circle* how the new realms revealed by the microscope and telescope influenced literature.

7. Hooke, *Posthumous Works*, pp. 120–121. Hooke continues: "All those infinitely infinite Radiations, which proceed from the whole Hemisphere of the Universe, and pass through the Area of the Superficies of the Pupil of the Eye, are by this truly wonderful Contrivance of the Eye, separated from each other, and conveyed to the distinct cells of the Microcosm of the Eye."

8. Hooke, *Posthumous Works*, pp. 76–77. On the collective effect of atoms, Lucretius wrote in his epic poem *De Rerum Natura* (Latham, p. 69): "Often on a hillside fleecy sheep, as they crop their lush pasture, creep slowly onward, lured this way or that by grass that sparkles with fresh dew, while the full-fed lambs gaily frisk and butt. And yet, when we gaze from a distance, we see only a blur—a white patch stationary on the green hillside. Take another example. Mighty legions, waging mimic war, are thronging the plain with their maneuvers. The dazzling sheen flashes to the sky and all around the earth is ablaze with bronze. Down below there sounds the tramp of a myriad marching feet. A noise of shouting strikes upon the hills and reverberates to the celestial vault. Wheeling horsemen gallop hot-foot across the midst of the plain, till it quakes under the fury of their charge. And yet there is a vantage-ground high among the hills from which all these appear immobile—a blaze of light stationary upon the plain."

9. From Fontenelle's *Plurality of Worlds*, translated by Joseph Glanville, pp. 122–124. Steven Dick in his *Plurality of Worlds* discusses the numerous editions and various translations of Fontenelle's *Entretiens sur la Pluralité des Mondes* (1686), and also of Huygens's *Cosmotheros* published posthumously in 1698 in Latin (in English the same year as *The Celestial Worlds Discover'd: or, Conjectures Concerning the Inhabitants of the Planets*), which covers much the same ground more technically. Marjorie Nicolson in "The early stages of Cartesianism in England" says that Fontenelle's *Plurality of Worlds*, translated a dozen times into English, made it fashionable to invite "philosophers from academic cloisters into the salons of ladies. The effect upon feminine society was particularly noteworthy; if the fair Marchioness of M. Fontenelle could understand Cartesianism, so could the ladies of England."

10. *The Celestial Worlds Discover'd*, p. 156. For a fuller treatment of Christiaan Huygen's work on light, see his *Treatise on Light* (1690).

6. Newton's Needles and Halley's Shells

1. The Stoic extramundane void, elaborated by Neoplatonists, developed by Thomas Bradwardine (1290?–1349), Francesco Patrizi (1529–1597), Giordano Bruno (1548–1600), Tommaso Campanella (1568–1639), Pierre Gassendi (1592–1655), and Henry More (1614–1687), became the space of the Newtonians. See Samuel Sambursky, *The Physical World of Late Antiquity*, Alexandre Koyré, *From the Closed World to the Infinite Universe*, and Edward Grant, "Medieval and seventeenth-century conceptions of an infinite void space" and *Much Ado about Nothing*. See Benjamin Brickman, "On physical space," which contains a translation of Patrizi's first book of the 1593 edition of *Pancosmia*, and Lillian Pancheri, "Pierre Gassendi, a forgotten but important man in the history of physics." See also Edwin Burtt, *Metaphysical Foundations of Modern Physical Science*, for discussions on Henry More.

2. This manuscript, written in a notebook and commencing with the words, *De gravitatione et aequipondio fluidorum*, is translated under the title "On the gravity and equilibrium of fluids" in *Unpublished Scientific Papers of Isaac Newton*, edited and translated by A. Rupert Hall and Marie Boas Hall, p. 121. The following interesting quotation is from Newton's essay, "Cosmography," pp. 375–376: "The Universe consists of three sorts of great bodies, Fixed Stars, Planets, & Comets, & all these have a gravitating power tending towards them by which their parts fall down to each of them after the same manner as stones & other parts of the Earth do here towards the earth ... The fixt Stars are very great round bodies shining strongly with their own heat & scattered at very great distances from one another throughout the whole heavens ... Our Sun is one of ye fixt Stars & every fixt star is a Sun in its proper region. For could we be removed as far from ye Sun as we are from ye fixt stars, the Sun by reason of its great distance would appear like one of ye fixt stars. And could we approach as neare to any of ye fixt Stars as we are to ye Sun, that Star by reason of its nearness would appear like our Sun ... For ye milky way being viewed through a good Telescope appears very full of very small fixt stars & is nothing else then ye confused light of these stars. And so ye fixt clouds & cloudy stars are nothing else then heaps of stars so small & close together that without a Telescope they are not seen apart, but appear blended together like a cloud."

3. Edmund Halley, "Philosophiae naturalis principia mathematica" (1687). First and foremost Newton was a mathematical genius. Less well-known is the fact that he was an ardent alchemist. From the 1660s to the early 1690s he devoted most of his time to transcribing an immense body of arcane literature and to performing in his laboratory numerous chemical experiments with care and skill. Also he was a lifelong, secretively heterodox, unitarian theologian. He wrote voluminously on the meaning of sacred texts, unraveled the apocalypse according to St. John and the prophecies in the Book of Daniel, studied Mosaic chronology and early church history, and in his posthumously published *Chronology of Ancient Kingdoms Amended* (1728), he reconstructed history and brought it into conformity with scriptural records. In *Foundations of Newton's*

Alchemy, Betty Dobbs writes that most of Newton's "great powers were poured out upon church history, theology, 'the chronology of ancient kingdoms,' prophecy, and alchemy."

4. Richard Bentley, "Eight Boyle Lectures," in vol. 3, *The Works of Richard Bentley,* edited by Alexander Dyce and published in 1838. The lectures bear the titles: (1) "The folly of atheism and (what is now called) deism"; (2) "Matter and motion cannot think: or a confutation of atheism from the faculties of the soul"; (3, 4, 5) "A confutation of atheism from the structure and origin of human bodies"; and (6, 7, 8) "A confutation of atheism from the origin and frame of the world." The seventh and eighth lectures are reproduced in *Isaac Newton's Papers and Letters on Natural Philosophy and Related Documents,* edited by I. Bernard Cohen and Robert Schofield. This latter work contains Perry Miller's "Bentley and Newton," p. 271. See also, Henry Guerlac and M. C. Jacob, "Bentley, Newton, and providence (the Boyle Lectures once more)."

5. *The Correspondence of Isaac Newton,* vol. 3, edited by H. W. Turnbull et al., pp. 233 (December 10, 1692), 238 (January 17, 1693), 244 (February 11, 1693), and 253 (February 25, 1693).

6. Lucretius, *De Rerum Natura* (Latham) p. 56.

7. See Michael Hoskin, "The English background to the cosmology of Wright and Herschel," and "Newton, providence, and the universe of stars." The quotation is from the second edition of *Principia* (edited by I. Bernard Cohen), vol. 2, p. 236.

8. *Isaac Newton's Mathematical Principles of Natural Philosophy and his System of the World,* 2nd ed., translated by Andrew Motte, introduced by I. Bernard Cohen, vol. 2, pp. 389–390.

9. Isaac Newton's *Opticks,* 4th ed., p. 400.

10. Voltaire, *Letters Concerning the English Nation* (1733).

11. Edmund Halley, "Considerations on the change of the latitude of some of the principal fixt stars" (1718).

12. The full title of this paper by Halley reads, "An account of several nebulae or lucid spots like clouds, lately discovered among the fixt stars by help of the telescope" (1714).

13. The first of these two papers was read before the Royal Society on March 9, 1721, and the second a week later on March 16, according to the Journal Book. See Michael Hoskin, "Dark skies and fixed stars."

14. Michael Hoskin, in "Dark skies and fixed stars," suggests that Halley might have heard the argument from David Gregory, Scottish astronomer, nephew of James Gregory, and professor of astronomy at Oxford. More recently, in "Stukeley's cosmology and the Newtonian origins of Olbers's paradox," Hoskin offers a different suggestion. William Stukeley (1687–1765), a minor figure in eighteenth-century science, in his *Memoirs of Sir Isaac Newton's Life* (1752), recalled that thirty years previously, somewhere between 1718 and 1721 and early in their acquaintance, he and Newton had a memorable conversation. He remembered how he discussed with Newton the question (p. 75): "What would have been the consequence had infinite space *quaquaversum* [in every direction]

been disseminated with worlds? We see every night the inconvenience of it. The whole hemisphere would have had the appearance of that luminous gloom of the Milky Way." Stukeley said also that he and Halley breakfasted with Newton on February 23, 1721. This suggests, says Hoskin, that Halley first heard the argument from Stukeley. Unfortunately, Stukeley's biography contains numerous factual errors and is largely responsible for the false image of Newton as an unassuming, tolerant person, untainted by heretical leanings and alchemical interests. Not impossibly, the fashions of thought and new ideas abroad in the mid-eighteenth century had influenced Stukeley's recollection of the content of the conversations held thirty years previously.

7. A Forest of Stars

1. Anecdotal details of Chéseaux's comet may be found in J. L. E. Dreyer, "The multiple tail of the great comet of 1744." See also Amédée Guillemin, *The World of Comets*, p. 212, and George Chambers, *The Story of the Comets*, pp. 128–129.

2. The two papers by Halley on the infinity of the universe discussing the darkness riddle were presumably well-known among astronomers. With other selected articles by different authors, they had been reproduced in a special six-volume edition of *Philosophical Transactions* (1734) of the Royal Society. Chéseaux's failure to acknowledge previous work may have misled recent writers. For example, Otto Struve in "Some thoughts on Olbers' paradox" (1963), Gustav Tammann in "Jean-Philippe de Loys de Chéseaux and his discovery of the so-called Olbers' paradox" (1966), and Donald Clayton in *The Dark Night Sky: A Personal Adventure in Cosmology* (1975), unaware of the earlier history of the riddle, attribute its origin to Chéseaux. For comments on this oversight see Stanley Jaki, "Olbers', Halley's, or whose paradox?" and Michael Hoskin's "Dark skies and fixed stars."

3. Chéseaux's calculation, updated and in the form given by Lord Kelvin in 1901, is as follows. We assume that all stars are sunlike, of radius a, and uniformly distributed with density n per unit volume. The number of stars in a shell of radius q and thickness dq equals $4\pi nq^2dq$, and the sum of their uneclipsed areas is this number multiplied πa^2, thus giving $4\pi^2na^2q^2dq$. If we divide this area of stellar disks by the area $4\pi q^2$ of the shell, we find that the fraction of the sky covered by the stars is πna^2dq. Let $\sigma = \pi a^2$ denote the geometric cross-section of a star. Hence the fraction of the sky covered by stars in the shell equals $n\sigma dq$. We now integrate out to distance r and find that the fraction of the sky covered is

$$\alpha = n\sigma r = r/\lambda$$

where $\lambda = 1/n\sigma$ is the mean free path of a light ray. The mean free path—a term commonly used in the kinetic theory of gases—is the average distance a particle travels between collisions. We note that α is unity when the radius r of a surrounding sphere of stars equals λ, and hence λ is also the background limit.

See E. Harrison, "Kelvin on an old and celebrated hypothesis." If $V = 1/n$ is the average volume occupied by one star, then

$$\lambda = V/\sigma$$

which is the equation in the text giving the background limit in a star-filled universe.

4. For a discussion of Gregory's photometric method of determining stellar distances, see Michael Hoskin, "Newton, providence and the universe of stars," and "The English background to the cosmology of Wright and Herschel." Albert Van Helden, *Measuring the Universe: Cosmic Dimensions from Aristarchus to Halley*, mentions James Gregory, p. 158.

5. We assume that all trees are alike, of width w at eye level, and uniformly distributed with density n per unit area. The number of trees in a zone of radius q and width dq equals $2\pi nqdq$, and the sum of their widths is $2\pi nqwdq$. If we divide this sum by the circumference $2\pi q$ of the zone, we obtain $nwdq$ for the tree-covered fraction. We integrate out to distance r and find:

$$\alpha = nwr = r/\lambda$$

where in this case $\lambda = 1/nw$ is the mean free path. We note that α equals unity when $r = \lambda$, and hence λ is the background distance. The average density of trees is $n = 1/A$, where A is the average area occupied by one tree, and we find

$$\lambda = A/w$$

which is the equation in the text giving the background limit in a forest. Alternatively, $\lambda = L^2/w$, when $A = L^2$, where L denotes a typical distance between trees. The number of trees within the background limit is hence

$$N = \pi\lambda^2/A = \pi A/w^2$$

and this is the number of visible trees given in the text. The forest analogy is treated in my article, "The dark night-sky paradox." Giordano Bruno was perhaps the first person to use the forest analogy. In *Jordani Bruni Nolani Opera Latine Conscripta*, vol. 1, 2nd pt., pp. 127–128, we find, "Look around in a forest in which the trees are planted everywhere with the same density. Would you credit that those nearby are separated from one another by distances greater than those far off, and that the distant confusion of many trees converts thousands into one?" As Stanley Jaki points out in his *Paradox of Olbers' Paradox*, by not drawing the logical conclusion from the forest analogy, Bruno failed to discover the riddle of darkness. Jaki's comments on the same score concerning Otto von Guericke, however, seem inapplicable because of Guericke's belief in the Stoic system.

6. The optical depth in a uniform medium is distance divided by the mean free path. Thus, if a cloud has an optical depth τ, it has a thickness of τ mean free paths. An osbcuring thing is said to be optically thick when its thickness exceeds one mean free path. This means it is larger than what we have called the background limit. The intensity of all forms of radiation—radio, infrared,

visible, ultraviolet, and x-ray—decreases according to exp-τ, where $\tau = r/\lambda$ denotes optical depth.

7. In note 3 the sky is covered with the stars out to the background limit $r = \lambda$. Let $N = 4\pi nr^3/3$ stand for the number of stars out to distance r; the number of stars needed to cover the sky is therefore

$$N = 4\pi n\lambda^3/3$$

If $V = 1/n$ represents the average volume occupied by a star, and $\lambda = V/\sigma$, then

$$N = 4\pi V^2/3\sigma^3$$

which is the equation in the text for the number of stars covering the sky.

8. The Misty Forest

1. This result can be derived formally as follows. We assume that all stars are similar to the Sun, of radius a, and are uniformly distributed with density n. The number of stars in a shell of radius q and thickness dq is $4\pi nq^2 dq$, and the fraction of the sky covered by the stars in the shell is this number multiplied by $a^2/4q^2$. In order to include the effect of geometric overlap by stars in intermediate shells, we multiply by $\exp(-q/\lambda)$, where $\lambda = 1/\pi na^2$ is the background limit. Furthermore, to include the effect of interstellar absorption, we multiply by $\exp(-q/\mu)$, where μ is the absorption mean free path. Hence, the fraction of the sky covered by the stars in the shell is $\lambda^{-1}\exp[(\lambda^{-1} + \mu^{-1})q]dq$. Integrating from $q = 0$ to $q = r$, we find

$$\alpha = \frac{\mu}{\mu+\lambda}\left(1 - \exp[(\lambda^{-1} + \mu^{-1})r]\right)$$

where α is the fraction of the sky covered by unobscured stars. As r goes to infinity in an unlimited and uniform universe, the fraction of the sky covered by stars becomes

$$\alpha = \mu/(\mu + \lambda)$$

If interstellar absorption is unimportant and the cut-off distance μ (which we have called the absorption limit) is much greater than the background limit λ, then α is unity, and the sky is fully covered with unobscured stars. If, however, interstellar absorption is important and the absorption limit μ is much less than the background limit λ, then $\alpha = \mu/\lambda$, and most stars are obscured from view. As stated in the text, the condition for darkness caused by an absorbing medium is that the absorption limit μ is much less than the background limit λ.

2. Olbers's article "Ueber die Durchsichtigkeit des Weltraumes" was published in 1823 in J. E. Bode's *Astronomical Year Book for the Year 1826*. A not-very-accurate translation in English, entitled "On the Transparency of Space," appeared in 1826 in the *Edinburgh New Philosophical Journal*.

3. See Otto Struve, "Some thoughts on Olbers' paradox"; also Stanley Jaki,

"Olbers', Halley's, or whose paradox?" and "New light on Olbers's dependence on Chéseaux."

4. John Herschel, "Humboldt's *Kosmos*," reproduced in *Essays*, p. 257. For comments on the conservation of energy and Herschel's rejection of absorption see Chapter 18.

5. Hermann Bondi, *Cosmology*, p. 21; also "Modern theories of cosmology."

6. Edward Fournier d'Albe, *Two New Worlds* (1907), p. 109.

9. Worlds on Worlds

1. Emanuel Swedenborg (1688–1772) of Stockholm, a Cartesian who turned to mysticism, wrote in 1734 in his *Principia Rerum Naturalium*, "The common axis of the sphere or starry heaven seems to be the galaxy, where we perceive the greatest number of stars. Along the galaxy all the vortices are in a rectilinear arrangement and series." English translation by James Rendel and Isaiah Tansley, vol. 2, p. 159. On the basis of vague statements of this kind, Swedenborg is often credited with being the first in the post-Newtonian era to theorize on galactic structure.

2. The full title of Wright's book is *An Original Theory or New Hypothesis of the Universe, Founded upon the Laws of Nature and Solving by Mathematical Principles the General Phaenomena of the Visible Creation; and Particularly the Via Lactea*. A facsimile reprint in 1971 contains an introduction by Michael Hoskin. Quotations are from pp. 57, 83–84. Thomas Wright's knowledge of astronomy came from popular and inspiring works such as Christiaan Huygens's *Celestial Worlds Discover'd* (1698), David Gregory's *Elements of Physical and Geometrical Astronomy* (1715), William Whiston's *Astronomical Lectures* (1715), John Keill's *Introduction to the True Astronomy* (1721), and William Derham's *Astrotheology, or a Demonstration of the Being and Attributes of God, from a Survey of the Heavens* (1715). See Michael Hoskin, "The Cosmology of Thomas Wright of Durham" (1970), and "The English Background to the Cosmology of Wright and Herschel" (1977).

3. In *Discoveries and Opinions of Galileo*, translated with an introduction and notes by Stillman Drake, p. 49.

4. Pierre Maupertuis (1698–1759), "Discours sur la figure des astres" (1742), in *Ouvres de Maupertuis* (1756), vol. 1. The passage referred to is quoted by Kant, *Natural History and Theory of the Heavens*, translated by W. Hastie, p. 62.

5. Lucretius, *De Rerum Natura* (Latham), p. 29.

6. The full title of Kant's book, translated by W. Hastie, reads *Universal Natural History and Theory of the Heavens; An Essay on the Constitution and Mechanical Origin of the Whole Universe Treated According to Newton's Principles*. The printing history of this work is given in Appendix C of a revised edition with an introduction by Willy Ley. In his introduction to *Kant's Cosmogony*, translated by W. Hastie, Gerald Whitrow says of Kant, "In the light of our current knowledge of the universe, it is clear that we should pay somewhat less attention than has been customary to those largely outmoded philosophical works of his later years

on which his reputation as a thinker has come to depend, and considerably more attention to those early scientific writings in which he revealed such remarkable physical insight." The review of Wright's *An Original Theory* in the Hamburg journal is reproduced as appendix B in W. Hastie's translation. The review quotes verbatim from Wright's book: "It only now remains to show how a number of stars, so disposed in a circular manner round any given center, may solve the phenomena before us. There are but two ways possible to be proposed by which it can be done, and one of which is highly probable; but which of the two will meet with your approbation I do not venture to determine. The first is in the manner I have described, i.e. all moving the same way, and not much deviating from some plane as the planets in their heliocentric motion do round the solar body . . . The second method of solving this phenomena is by a spherical order of the stars, all moving with different directions round one common center, as the planets and comets do round the Sun, but in a kind of shell or concave order. The former is easily conceived from what has already been said, and the latter is as easy to be understood if you have any idea of the segment of a sphere." Hoskin argues in his introduction to *An Original Theory* that Kant "creatively misunderstood" Wright when he adopted the disk system. Gerald Whitrow disagrees in his introduction to *Kant's Cosmogony* and remarks that even though Wright "seems to have hankered after a spherical interpretation of the stellar system," nonetheless in the crucial passage quoted above, which was in the Hamburg journal, Wright's perfectly clear description of a flat ring-shaped system was consistent with a disk model and lent itself to no misunderstanding. Quotations are from pp. 65 and 145 of Kant's *Universal Natural History,* translated by W. Hastie.

7. William Herschel was the first to identify Uranus as a planet and not a star. It had been observed many times before, and was included in a star map prepared by John Flamsteed, the first astronomer royal and compiler of important star catalogs.

8. See William Herschel, "On the construction of the heavens" (1785); also Michael Hoskin, *William Herschel and the Construction of the Heavens,* p. 99.

9. Quoted in Agnes Clerke, *The Herschels and Modern Astronomy* (1901), p. 67. "They were what Lambert and Kant had supposed them to be," she wrote (p. 66), "island universes, vast congeries of suns, independently organized, and of galactic rank. They were, each and all, glorious systems, barely escaping total submergence in the illimitable ocean of space." Alexander von Humboldt popularized the term "Weltinsel" or "island universe" in *Cosmos* (vol. I, p. 88, translated by E. C. Otté, 1855): "The cluster of stars, to which our cosmical island [Weltinsel] belongs, forms a lens-shaped, flattened stratum, detached on every side, whose major axis is estimated at seven or eight hundred and its minor one at a hundred and fifty times the distance to Sirius." The terms "island universe" and "island universes," the first implying a Stoic system and the second a Wright-Kantian-Epicurean system, should not be confused. To avoid this possibility, I distinguish between the island universe, referred to as the one-island or the one-galaxy universe, and the island universes, referred to as the many-

island or the many-galaxy universe. Richard Proctor in *Our Place among the Infinities* (pp. 193–194) commented on the change in Herschel's cosmological ideas: "As the work progressed Sir William Herschel grew less confident. He began to recognize signs of a complexity of structure which set his method of star-gauging at defiance. It became more and more clear to him also, as he extended his survey, that the star-depths were in fact unfathomable."

10. Revelations of Chaos

1. See *Outlines of Astronomy*, sec. 871.

2. Agnes Clerke, *The System of the Stars*, p. 368.

3. Friedrich Struve, *Etudes d'Astronomie Stellaire sur la Voi Lactée et sur les Distances des Étoiles Fixes* (1847). According to Struve, the Milky Way consists of a flattened system of stars that extends to indefinite distance. We cannot see the most distant stars in the Milky Way, he said, because of the extinction of starlight by absorption. Struve's picture comes reasonably close to the modern picture. The disk of the Galaxy, however, is not indefinitely large, as he said, but has a diameter of about 100,000 light-years. Absorption by gas and dust blocks our line of sight in the plane of the disk, and the Milky Way is not large enough and could not, even without absorption, blaze as brightly as the surface of the Sun.

4. The term "new astronomy" was used for the first time by the American astronomer Samuel Langley in his book *The New Astronomy* (1888). Alfred Tennyson, poet laureate and visionary spokesman of the Victorian era wrote in "Locksley Hall Sixty Years After" (1889):

Warless? when her tens are thousands, and her thousands millions, then—
All her harvests all too narrow—who can fancy warless men?

Warless? war will die out late then. Will it ever? late or soon?
Can it, till this outworn earth be dead as yon dead world the moon?

Dead the new astronomy calls her.—On this day and at this hour,
In this gap between the sandhills, whence you see the Locksley tower.

Quotations in the text, unless otherwise indicated, are from William Huggins, "The new astronomy: a personal retrospect," in *The Nineteenth Century* (1897). For further references see William and Margaret Huggins, *An Atlas of Representative Stellar Spectra* (1899) and *The Scientific Papers of Sir William Huggins* (1909); see also *A Sketch of the Life of Sir William Huggins* (1936), edited by Charles Mills and C. F. Brooks.

5. August Comte, in *Cours de Philosophie Positive* (1830–1842), 19th lecture, said, "Any research which cannot be reduced to actual visual observation is excluded where the stars are concerned . . . We can see the possibility of determining their forms, their distances, their magnitudes, and their movements, but it is inconceivable that we should ever be able to study by any means whatsoever their chemical composition or mineralogical structure, still less the nature

of the organic bodies living on their surface, etc." See *The Essential Comte*, p. 74.

6. William Huggins, "On the spectrum of the Great Nebula in Orion" (1865).

7. E. N. da C. Andrade, "Doppler and the Doppler effect"; Karel Hujer, "Sesquicentennial of Christian Doppler."

8. William Huggins, "Further observations on the spectra of stars and nebulae, with an attempt to determine therefrom whether these bodies are moving towards or from the Earth" (1868). Huggins acknowledged Doppler's idea, and in his "New astronomy" essay gave an illuminating analogy: "To a swimmer striking out from the shore each wave is shorter, and the number he goes through in a given time is greater than would be the case if he stood still in the water."

9. Robert Richardson, *The Star Lovers*, p. 158, gives the name of the magazine containing the article on how to make your own spectroscope as *Good Words*.

10. See Otto Struve and Velta Zebergs, *Astronomy of the 20th Century*, chaps. 19 and 20; also Michael Hoskin, "The 'Great Debate': what really happened." In the public discussion in 1920, Shapley, by studying the distribution of globular clusters, had succeeded in constructing for the first time a realistic picture of the size and shape of the Galaxy; his insufficient allowance for the extinction of starlight, however, had the effect of overestimating the size of the Galaxy. Curtis, on the other hand, underestimated the size of the Galaxy. Shapley in *Through Rugged Ways to the Stars* (p. 79), wrote in 1969: "As for the actual 'debate,' I must point out that I had forgotten about the whole thing long ago, and nobody had mentioned it to me for many years." It might be said that the long drama of the Stoic versus Epicurean systems climaxing in the controversy of the one-island versus many-island universe has been rather trivialized by limiting the account to the debate of 1920.

11. Charles Whitney, recalling his experiences as Shapley's graduate student at Harvard, writes in *The Discovery of Our Galaxy*, p. 219: "I have never seen a quicker mind, a more agile sense of humor, or a more complete absence of what usually passes for humility."

12. Shapley, "Colors and magnitudes in stellar clusters. Second part: Thirteen hundred stars in the Hercules Cluster (Messier 13)," p. 139, footnote. The view expressed by Shapley had been widely shared in previous decades. Simon Newcomb, a Canadian-born astronomer who rose to the rank of rear admiral at the Naval Observatory in Washington, in his book *The Stars: A Study of the Universe* (1901), pp. 231–233, discussed various solutions to the riddle of darkness and said he favored a finite material system (a one-galaxy universe) surrounded by infinite and empty space. "The hypothesis of a limited universe and no extinction of light, while not absolutely proved, must be regarded as the one to be accepted until further investigations shall prove its unsoundness."

13. Edwin Hubble, *The Realm of the Nebulae* (1936).

14. See Owen Gingerich, "Charles Messier and his catalog," and Kenneth Jones, *Messier's Nebulae and Star Clusters*.

11. The Fractal Universe

1. Kant, *Universal Natural History,* translated by W. Hastie, p. 65.

2. Johann Lambert, *Cosmological Letters on the Arrangement of the World-Edifice,* translated with an introduction and notes by Stanley Jaki. In a letter to Kant dated 1765, Lambert wrote: "I can tell you with confidence, dear sir, that your ideas about the origin of the world, which you mention in the preface to *Only Possible Proof* were not known to me before. What I said on page 149 of *Cosmological Letters* dates from 1749. Right after supper I went to my room, contrary to my habit then, and from my window I looked at the starry sky, especially the Milky Way, I wrote down on a quarto sheet the idea that occurred to me then that the Milky Way could be viewed as an ecliptic of the fixed stars, and it was this note I had before me when I wrote the *Letters* in 1760."

3. John Herschel reviewed the first volume of "Humboldt's *Kosmos"* in 1848 in the *Edinburgh Review.* He wrote (p. 98), "The assumption that the extent of the starry firmament is literally infinite has been made by one of the greatest astronomers, the late Dr. Olbers, the basis of a conclusion that the celestial spaces are in some slight degree deficient in *transparency;* so that all beyond a certain distance is, and must forever remain, unseen; the geometrical progression of the extinction far outrunning the effect of any conceivable increase in the power of our telescopes. Were it not so, it is argued, every part of the celestial concave ought to shine with the brightness of the solar disc; since no visual ray could be so directed as not, in some point or other of its infinite length, to encounter such a disc. With this peculiar form of the argument we have little concern. It appears to us, indeed, with all deference to so high an authority, invalid." John Herschel then suggested the hierarchical solution, although he himself, preferring a one-galaxy universe, believed in the Stoic solution. Richard Proctor, *Other Worlds than Ours* (1871), in a footnote, p. 289, quotes from a letter dated August 20, 1869, in which John Herschel expresses his views. Similar ideas had occurred to Proctor while writing a series of essays entitled "A new theory of the universe" in 1869 for the *Student.* Excerpts from the Herschel–Proctor correspondence may be found in *Knowledge,* pp. 83–85, January 1, 1886. We note that the word "hierarchy" comes from the Greek *hier,* meaning sacred, and *archy,* meaning rule. We speak properly of a hierarchy of angels, or priests, or bureaucrats, but improperly of a hierarchy of physical objects. More correctly, we should speak of a multilevel or multilayer universe, not a hierarchical universe. But we need not be too fussy about using the word "hierarchy" in a catch-all capacity; like other words, it means whatever fashion decrees. See the preface to *Hierarchical Structures,* edited by Lancelot Whyte, Albert Wilson, and Donna Wilson. Gerard de Vaucouleurs discusses the modern evidence for a hierarchical universe in "The case for a hierarchical cosmology."

4. Benoit Mandelbrot, *The Fractal Geometry of Nature.*

5. Woods are never exactly circular, and the numerical coefficient is not important. In any case, averaging the optical depth over all directions in a cluster adjusts the coefficient to unity, as assumed in the text.

6. The condition derived in the text for a transparent or optically thin forest is independent of the number of levels of clustering, provided all clusters are individually transparent. An alternative determination is the following. Let L_i be the diameter of an ith-level cluster composed of N_i subclusters of diameter L_{i-1}. An observer sees out of a cluster of diameter L_i between N_i subclusters, each of diameter L_{i-1}, when the background limit $L_i^2/N_i L_{i-1}$ due to the subclusters exceeds L_i, or

$$N_i < L_i/L_{i-1}$$

This condition, satisfied at all levels, provides a sufficient condition for a transparent forest. We note that the total number of trees in the ith-cluster is $N = N_1 N_2 N_3 \ldots N_{i-1} N_i$ and hence

$$N < L_i/L_0$$

where L_i is the diameter of the cluster containing N trees and L_0 is the width of a tree. This is the condition derived in the text. More generally, let L denote the characteristic size of a region in space and N denote the total number of members in that region. In self-similar clustering, N varies as L^D, where D is the fractal dimension. A uniform distribution has a fractal dimension of $D = d$ in a space of d dimensions. Two fractals can coexist without overlap in a space of d dimensions when the sum of their fractal dimensions is less than d. The fractal dimension of a ray of light is unity. A structure is optically thin and hence coexists with rays of light when the sum of fractal dimensions $D + 1$ is less than d. A forest of $d = 2$ is optically thin when its fractal dimension D is less than 1. Notice that the average density of the members of a fractal is proportional to N/L^d, and this diminishes with size as L^{D-d}. A fractal of infinite extent has an average density of zero when $D < d$, even though its members are infinite in number for $D > 0$.

7. Averaging the optical depth over all directions in a spherical cluster has the same effect on the numerical coefficient mentioned in note 5.

8. The Seeliger–Charlier condition (see note 10, below) is derived as follows: An observer sees out of a cluster of diameter L_i between N_i subclusters each of diameter L_{i-1} when the background limit $L_i^3/N_i L_{i-1}^2$ due to the subclusters exceeds L_i or

$$N_i < (L_i/L_{i-1})^2$$

When satisfied at all levels, this relation is a sufficient condition for a transparent universe. If $N = N_1 N_2 N_3 \ldots N_i$ is the total number of stars in the ith cluster, then

$$N < (L_i/L_0)^2$$

where L_0 is the diameter of a star. This is the condition derived in the text. For a configuration of fractal dimension D to be transparent we require $D + 1$ to be less than d, or D less than 2 in a 3-dimensional space. The average density of stars diminishes as L^{D-3} and approaches zero in an optically thin hierarchical universe of infinite extent.

9. Edward Fournier d'Albe, *Two New Worlds*. Quotations are from pp. 98, 100.

10. Carl Charlier's first publication on hierarchy, "Wie eine unendliche Welt aufgebaut sein kann" (1908), contains mathematical errors. The correct result, given in note 8 above, was first derived in 1909 by Hugo von Seeliger in a letter to Charlier, who acknowledged and used this result in a second article, "How an infinite world may be built up" (1922). We note that the gravitational potential at the center of a cluster of mass M and size L is proportional to L^{D-1}, and hence astronomical hierarchies of D less than 1 are exempt from the Dirichilet problem of an indefinite or infinite potential. Automatically the sky is dark for a hierarchical universe in which the gravitational potentials and equilibrium speeds of clusters diminish at higher levels.

11. Svante Arrhenius, "Infinity of the universe" (1911), translated from German by J. F. Fries.

12. The Visible Universe

1. For discussions on the visual-ray and eidola theories see Joseph Priestley, *History and Present State of Discoveries Relating to Vision, Light, and Colours* (1772); John Herschel, *Light and Vision* (1831); Lynn Thorndike, *A History of Magic and Experimental Science;* Vasco Ronchi, *The Nature of Light;* and *A Source Book of Medieval Science*, edited by David Lindberg.

2. Galileo, *Dialogue Concerning Two New Sciences*, pp. 43–44. Galileo wrote in connection with this experiment, "What a sea we are gradually slipping into without knowing it! With vacua and infinities and indivisibles and instantaneous motions, shall we ever be able, even by means of a thousand discussions, to reach dry land?"

3. See A. L. Sabra, *Theories of Light from Descartes to Newton;* chap. 2, "Descartes' doctrine of the instantaneous propagation of light and his explanation of the rainbow and the colours," contains a summary of pre-Cartesian ideas on the speed of light.

4. Hooke, *Micrographia*, pp. 56–57.

5. Hooke, "Lectures of light, explicating its nature, properties, and effects, etc.," in *The Posthumous Works of Robert Hooke*, pp. 77–78.

6. See "Roemer and the first determination of the velocity of light" by I. Bernard Cohen. This paper gives an analysis of Roemer's method and data, a bibliography, and an English translation of Roemer's original paper: "A demonstration concerning the motion of light." Cohen summarizes the pre-Roemer ideas concerning the finite speed of light. Newton knew from Halley's review of the data on Io's eclipses in 1694 that light travels an astronomical unit in 8 minutes 30 seconds, which happens to be only 10 seconds more than the present-day determination. The value of the astronomical unit was not known with the same precision. At that time neither Newton nor anybody else had reason for using light-travel time as a measurement of distance.

7. Francis Roberts, "Concerning the distances of the fixed stars."

8. George Sarton, "Discovery of the aberration of light." Bradley's discovery was communicated to Halley and published under the title, "A letter from the Reverend Mr. James Bradley, Savilian Professor of Astronomy at Oxford, and F.R.S., to Dr. Edmond Halley Astronom. Reg. &c, giving an account of a new discovered motion of the fix'd stars." Sarton's paper includes a reproduction of Bradley's letter to Halley. Jean Picard, Robert Hooke, and John Flamsteed had previously observed annual periodic displacements of stars caused by aberration. Hooke and Flamsteed mistakenly claimed that they had detected parallax, but Jean Cassini in 1699 pointed out that their observations were inconsistent with annual parallax.

9. Thomas Young (1773–1829), scientist, physician, authority on Egyptian hieroglyphics, established the wave theory of light and founded modern studies of color vision. Subsequent developments in the wave theory of light were made by Augustin Fresnel in France, and in color vision by Hermann von Helmholtz in Germany.

10. Agnes Clerke, *The System of the Stars* (1890), pp. 380–381. "For our view of sidereal objects is not simultaneous. Communication with them by means of light takes time, and postdates the sensible impressions by which we are informed of their whereabouts in the direct proportion of their distances. We see the stars not where they are—not even where they were at any one instant, but where they were on a sliding scale of instants."

11. Quoted from William Herschel's preface in the "Catalogue of 500 new nebulae, nebulous stars, planetary nebulae, and clusters of stars; with remarks on the construction of the heavens" (1802).

12. John Herschel, *Treatise of Astronomy,* p. 354. This work was elaborated in 1849 into the famous *Outlines of Astronomy* that ran through twelve editions—the last in 1873—and became the most widely read learned astronomy text of the nineteenth century. Herschel estimated that our sidereal system (Galaxy) contains 100 million stars and extends to a distance of 2,000 light-years.

13. John Pringle Nichol's popular *Views of the Architecture of the Heavens* (1838) concludes with several informative essays in the form of notes. This book brought the developments in astronomy to the attention of the English-speaking public and influenced Edgar Allan Poe. See "Lord Kelvin and his first teacher in natural philosophy"; Silvanus Thompson, *The Life of William Kelvin,* vol. 1, chap. 1; also Frederick Connor, "Poe and John Nichol."

14. Thomas Dick, born in Dundee in 1774, though remembered in theological circles, has been forgotten by astronomers despite the popularity of his astronomical books. He studied at Edinburgh, taught at school in Methven and Perth for twenty years until 1827, and thereafter devoted himself entirely to writing. His best-known scientific books were *Celestial Scenery* (1838) and *The Sidereal Heavens* (1840). The popular *Complete Works of Thomas Dick,* "illustrated with engravings and a portrait of the author," ran through numerous editions, particularly in America.

15. Alexander von Humboldt, *Cosmos: A Sketch of a Physical Description of the*

Universe, translated by E. C. Otté, 5 vols. (1848–1865) 88), vol. 1, pp. 144–145. See also vol. 3, pp. 105–106, for a discussion of the speed of light.

16. Richard Proctor, *The Expanse of Heaven: A Series of Essays on the Wonders of the Firmament* (1874), p. 203. A few pages later (206–287) he wrote, "But the star depths are never revealed to us exactly as they are at the moment, or exactly as they were at any moment." He considered the possibility of creatures sensitive to gravitational signals, for whom "the information conveyed respecting the universe would be far more nearly contemporaneous, since the action of gravity certainly travels many thousands of times faster than light, even if it do not travel with infinite velocity as some philosophers suppose." In this he erred, for gravity propagates at the speed of light, as was shown fifty years later by Albert Einstein.

17. Robert Ball, Irish astronomer, studied at Trinity College, Dublin, became royal astronomer for Ireland in 1874 and Lowndean professor of astronomy at the University of Cambridge in 1892. For the quotation see his book *In Starry Realms,* pp. 259–260.

13. The Golden Walls of Edgar Allan Poe

1. Edgar Allan Poe, "The Power of Words," first published in *United States Magazine and Democratic Review* (June 1845), reproduced in *The Science Fiction of Edgar Allan Poe* (1976), edited by Harold Beaver, p. 171.

2. William Browne, "Poe's *Eureka* and some recent scientific speculations" (1869); Frederick Connor, "Poe and John Nichol: notes on a source of *Eureka,*" in *All These to Teach: Essays in Honor of C. A. Robertson* (1949), edited by Robert Bryan et al.; and *Poe as Literary Cosmologer* (1975), edited by Richard Benton.

3. Poe, *Eureka: A Prose Poem* (1848), reproduced with a biographical guide by Richard Benton, and in *The Science Fiction of Edgar Allan Poe* (1976) edited by Harold Beaver. For this "pamphlet," dedicated to Alexander von Humboldt, and amplifying a two-hour lecture, "On the cosmogony of the universe," at the Society Library, New York, Poe received fourteen dollars from the publisher George Putnam. Only 500 copies were printed, which sold slowly, despite warm reviews in journals such as the *Weekly Universe.* Poe wrote in the preface, *"What I here propound is true*—therefore it cannot die—or if by any means it be now trodden down so that it die, it will 'rise again to the Life Everlasting.' Nevertheless it is as a Poem only that I wish this work to be judged after I am dead." Our main interest in *Eureka* relates to Poe's solution of the riddle of darkness, in which he proposed that the light from very distant stars has not yet reached us. This solution remains quite general and requires slight amendment in the modern expanding universe. The quotations from *Eureka* are from Benton's reproduction with the original pagination: pp. 100, 101–102, 117, and 136.

4. Letter to George Eveleth, February 29, 1848. See *The Science Fiction of Edgar Allan Poe,* edited by Harold Beaver, n. 1, p. 395.

5. Fournier d'Albe, *Two New Worlds,* p. 94.

6. Ibid., p. 95.

14. Lord Kelvin Sees the Light

1. See Silvanus Thompson, *The Life of William Thomson, Baron Kelvin of Largs*, vol. 1, chap. 1, "Upbringing at Glasgow." At a ceremony in memory of John Pringle Nichol, reported in "Lord Kelvin and his first teacher in natural philosophy" (1903), Kelvin recalled, "The benefit we had from coming under his inspiring influence, that creative influence, that creative imagination, that power which makes structures of splendour and beauty out of the material of bare dry knowledge, cannot be overestimated."

2. These handwritten notes by Hathaway were reproduced by papyrograph. The lectures were also summarized by George Forbes in "Molecular dynamics," *Nature* 32 (March 19, 1885): 461–463; (April 2) 508–510; and (April 30) 601–603.

3. The full title reads, *Baltimore Lectures on Molecular Dynamics and the Wave Theory of Light, Founded on Mr. A. S. Hathaway's Stenographic Report of Twenty Lectures Delivered in The Johns Hopkins University, Baltimore, October 1884; Followed by Twelve Appendices on Allied Subjects*. See Lecture 16, pp. 260–278. Following Kelvin's death in 1907 at age eighty-three, Joseph Larmor in an obituary, "William Thomson, Baron Kelvin of Largs," wrote that Kelvin's "purely scientific activity from 1884 onwards hinged largely on the production of the definitive edition of these lectures."

4. Kelvin referred to his ideas in a discourse on "The absolute amount of gravitational matter in any large volume of interstellar space," delivered at a British Association meeting in October 1901, and this discourse was summarized in *Nature* in an article entitled "On the clustering of gravitational matter in any part of the universe" (1901). Also, the salient points of the important 1901 *Philosophical Magazine* paper are stated again in 1902 in a second *Philosophical Magazine* paper bearing the same title as the article in *Nature*, and this second paper is reproduced as Appendix D in the *Baltimore Lectures*. In this second paper Kelvin repeats his remarks on the distribution of matter in the Galaxy and on how it affects the velocities of stars; he discussed gravitational free-fall collapse of a spherical body, and incidentally referred to his previous conclusions concerning the darkness of the night sky. This interesting paper on gravitational collapse is discussed in my "Newton and the infinite universe" (1986); though reproduced in Kelvin's *Mathematical and Physical Papers* and listed in Thompson's bibliography, it suffered a fate similar to his 1901 paper and remained neglected.

5. The following analysis parallels that by Kelvin in section eighteen of the revised Lecture 16. We assume for computational convenience that all stars are sunlike, of radius a, and uniformly distributed. Let n be the number per unit volume. Then the number in a shell of radius q and thickness dq equals $4\pi n q^2 dq$, and the fraction of the sky covered by the stars in the shell is this number multiplied by $\pi a^2 / 4\pi q^2$. To include geometric overlap (neglected by Kelvin), we multiply by $\exp(-q/\lambda)$, where $\lambda = 1/\pi n a^2$ is the mean free path of a light ray. The fraction of the sky covered by stars in the shell is hence $\lambda^{-1} \exp(-q/\lambda) dq$. By integrating from $q = 0$ to $q = r$ we find

$$\alpha = 1 - e^{-r/\lambda} \tag{1}$$

where α denotes the fraction of the sky covered by the disks of stars out to radius r. The entire sky is covered ($\alpha = 1$) in a distribution of stars of infinite extent and the average distance seen from any position is the background limit λ. We arrive at Kelvin's result when the radius r of the system of stars is small compared with the background limit:

$$\alpha = \frac{r}{\lambda} = \frac{3N}{4}\left(\frac{a}{r}\right)^2 \tag{2}$$

where $N = 4\pi n r^3/3$ is the total number of stars. This treatment by Kelvin elucidates that previously given by Halley in 1721, Chéseaux in 1744, and Olbers in 1823. Let each star have luminosity L. The contribution to the radiation density u at the center of the shell is $du = (nL/c)e^{-q/\lambda}dq$. By integrating as before, we find

$$u = u*(1 - e^{-r/\lambda}) \tag{3}$$

where $u* = L/\pi a^2 c$ is the radiation density at the surface of a star. Hence we find $u = u*$ in any distribution of stars extending beyond the background distance. From equations (1) and (3) we obtain

$$\alpha = u/u* \tag{4}$$

thus demonstrating the truth of Kelvin's statement that "α is the ratio of the apparent brightness of our star-lit sky to the brightness of our sun's disc." In discussions of the riddle in recent years this relation between sky-coverage and brightness has received scant recognition. This may be the result of neglecting Olbers's line-of-sight argument. See my paper "Kelvin on an old and celebrated hypothesis" (1986).

6. Simon Newcomb, in *Popular Astronomy* (1883, first edition 1878), presented the standard cosmological model of the nineteenth-century: a single-galaxy universe, the Milky Way, which consisted of 10^9 stars and had a radius of 1,000 parsecs, thus giving a typical separating distance of about 5 light-years between neighboring stars. See Agnes Clerke's *The System of the Stars* (1890), chap. 24, and in particular p. 368.

7. The horizon at the edge of the visible universe is usually referred to as the "particle horizon." Cosmological horizons are discussed by Wolfgang Rindler in "Visual horizons in world-models" (1956) and at a more elementary level by me in *Cosmology: The Science of the Universe* (1981), chap. 19.

8. William Thomson, "On mechanical antecedents of motion, heat, and light" (1854), reproduced in *Mathematical and Physical Papers*, vol. 2, p. 34.

9. See Joe Burchfield, *Lord Kelvin and the Age of the Earth* (1975). See also Kelvin's *Mathematical and Physical Papers*, 6 vols., and *Popular Lectures and Addresses*, 3 vols., for various discussions on the age of the Earth and the Sun.

10. William Thomson, "On the age of the Sun's heat" (1962), reproduced in *Popular Lectures and Addresses*, vol. 1, p. 349.

11. The background limit varies as L^3, where L is the average separating dis-

tance between stars. Reducing the background limit from 10^{23} to 10^{10} light-years requires that we multiply the present value of L by 5×10^{-5}. Taking the present value of L as 100 light-years (remembering that galaxies are widely separated), we obtain an average separating distance of 5×10^{-3} light-years, or 3,000 astronomical units. We should note in the case of galaxies, each occupying an average volume $V = 100$ and having a cross-section $\sigma = 10^{-2}$, measured in million-light-year units, that the background limit occurs at 10^{10} light-years. Hence, very roughly, galaxies in the visible universe out to a distance 10^{10} light-years cover the sky. But, of course, they fall to create a bright starlit sky. See W. Bonnor, "On Olbers' Paradox."

12. Normal stars cannot exist in a bright-sky universe and our equations should therefore be used to determine the conditions for dark skies.

13. Kelvin here considered the improbable situation, which I have discussed in "Why the sky is dark at night" (1974), of distributing luminous stars on the observer's backward lightcone. Let us suppose that all stars commence shining at time $t = 0$. Initially an observer sees a sphere of stars (the visible universe) radially expanding at the speed of light. After time t^*, denoting a typical stellar luminous lifetime, the nearby stars are extinguished and the observer is then surrounded by a radially expanding sphere of dark stars. Immediately outside this sphere of dark stars lies a receding shell of thickness ct^* containing luminous stars. Beyond the shell lie all the stars whose light has not yet reached the observer. We could argue that stars shine over many generations with each generation lasting 10 billion years. But if the matter content of the universe is conserved, then fewer stars shine in each generation, and the final intensity of light in the universe remains independent of the number of generations. Let us now suppose, with Kelvin, that stars commence shining at time $t = -r/c$, where r denotes the distance of each star. All stars now lie on the observer's backward lightcone and appear to be luminous simultaneously for a time $0 \leq t \leq t^*$. In this way the sky becomes covered with stellar disks. The observer in this improbable model has special location, contrary to the spirit of modern cosmology.

15. Ether Voids, Curved Space, and a Midnight Sun

1. Lord Kelvin, "Note on the possible density of the luminiferous medium, and on the mechanical value of a cubic mile of sunlight" (1854); reproduced with comments in *Notes of Lectures on Molecular Dynamics and the Wave Theory of Light* (1884), delivered at Johns Hopkins University. Kelvin's first paper in 1847 bore the title "On a mechanical representation of electric, magnetic, and galvanic forces." For essays on the history of the ether idea, see *Conceptions of Ether*, edited by G. N. Cantor and M. J. S. Hodge (1981).

2. Simon Newcomb, *Popular Astronomy* (1878), p. 505: the return of the heat rays of the Sun "can result only from space having such a curvature that what seems to us a straight line shall return into itself, as has been imagined by a great German mathematician; or from the ethereal medium, the vibrations in which constitute heat, being limited in extent; or, finally, through some agency

as yet totally unknown to science." In a footnote he identified the mathematician as Riemann.

3. John Ellard Gore (1845–1918), "On the infinity of space," chap. 27 in *Planetary and Stellar Studies* (1888), p. 233.

4. Gore failed to realize that the riddle of cosmic darkness cannot be solved by surrounding stars with reflecting walls. This can be very easily demonstrated by simple calculations. Regrettably, the history of the riddle of darkness at night consists of numerous discussions and proposed solutions, but very few calculations.

5. Lord Kelvin, "On the clustering of gravitational matter in any part of the universe" (1902).

6. Fournier d'Albe, *Two New Worlds* (1907), p. 100.

7. Simon Newcomb, "Elementary theorems relating to the geometry of a space of three dimensions and of uniform positive curvature in the fourth dimension" (1877). Mathematicians in the nineteenth century discovered three uniform spaces: Euclidean (zero curvature), hyperbolic (negative curvature), and spherical (positive curvature). Through any point in Euclidean space can be drawn only one straight line parallel to another straight line; through any point in hyperbolic space can be drawn an infinity of straight lines parallel to another straight line; and through any point in spherical space can be drawn no straight lines parallel to another straight line. Simon Newcomb, in *The Stars: A Study of the Universe* (1901), p. 226, commented that the "problem of the structure and duration of the universe is the most far-reaching with which the mind has to deal."

8. W. Barrett Frankland, "Notes on the parallel axiom" (1913). See John North's discussion of a closed universe in *The Measure of the Universe*, pp. 72–81; also Dionys Burger's treatment in *Sphereland*.

9. Willem de Sitter, "On Einstein's theory of gravitation and its astronomical consequences. Third paper," p. 25.

10. de Sitter, "On the curvature of space," footnote, p. 234.

11. Johann Zöllner, "Ueber die Endlichkeit der Materie im unendliche Raume," in *Über die Natur der Cometen* (1883), pp. 90–104. Stanley Jaki, who refers to this work in *The Paradox of Olbers' Paradox*, also advocates the finite-universe solution. Mathematical analysis shows, however, that the curvature of homogeneous and isotropic space has no effect on the radiation level. This is explained in my "Dark night sky paradox" (1977), and in the more technical "Radiation in homogeneous and isotropic models of the universe" (1977).

16. The Expanding Universe

1. Of 41 galaxies studied, Slipher found that 36 were receding and 5 approaching. Galaxies tend to cluster in groups of various sizes and have random motions superposed on the cosmic recession. Those nearby recede and approach with equal probability, those farther away mostly recede, and those even

more distant recede entirely. The Andromeda nebula, M 31, was the first galaxy to have its velocity measured. Slipher found in 1912 that it had a velocity of approach of about 300 kilometers per second. For a history of the developments early in the twentieth century see John North, *Measure of the Universe* (1965). See also William McCrea, "Willem de Sitter, 1872–1934" (1972). The best biographies of Einstein include Philipp Frank, *Einstein: His Life and Times* (1947); Banesh Hoffmann, *Albert Einstein: Creator and Rebel;* and Jeremy Bernstein, *Einstein* (1973).

2. Einstein's paper "Cosmological consideration of the general theory of relativity" is translated into English in *The Principle of Relativity: A Collection of Original Memoirs on the Special and General Theory of Relativity* by H. A. Lorentz, A. Einstein, H. Minkowski, and H. Weyl. We note here that special relativity applies in the absence of gravity. According to this theory, spacetime decomposes into time and space, and the space of each observer has Euclidean geometry. General relativity has similar metrical structure with Riemannian geometry, and applies more generally in the presence of gravity. An observer falling freely in a gravitational field experiences no gravity (is weightless) and uses special relativity to explain local happenings. The spherical space—and elliptical variant—of Einstein's cosmological model had been studied by Riemann and Newcomb. Also, Karl Schwarzschild, a student of Hugo von Seeliger and distinguished astronomer and mathematician, attempted in a paper "On the admissable curvature of space" (1900) to determine the curvature of space on the basis of parallax statistics. Einstein said in 1954 that Arthur Eddington's classic *Mathematical Theory of Relativity* (1923) was the finest presentation of the subject in any language. See A. Vibert Douglas, *Arthur Stanley Eddington* (1956).

3. The third and, for our purpose, most important of de Sitter's papers, "On Einstein's theory of gravitation and its astronomical consequences," was published in 1917. In contrast to Einstein's static model, called hypothesis A, de Sitter proposed a dynamic model, called hypothesis B. Both models used the cosmological term introduced by Einstein that acts in opposition to gravity. In his discussion of the relative merits of the two models, de Sitter wrote, "If, however, continued observation should confirm the fact that the spiral nebulae have systematically positive radial velocities, this would certainly be an indication to adopt the hypothesis B in preference to A. If it should turn out that no such systematic displacement of spectral lines towards the red exists, this could be interpreted either as showing A to be preferable to B, or as indicating a still larger value of R in the system B." In de Sitter's paper R equals $(3/\Lambda)^{1/2}$, and Λ denotes the cosmological term.

4. Arthur Eddington, *The Mathematical Theory of Relativity* (1923), p. 162.

5. See Aleksandr Friedmann, "On the curvature of space" (1922); Georges Lemaître, "A homogeneous universe of constant mass and increasing radius accounting for the radial velocity of extra-galactic nebulae" (1931).

6. Edwin Hubble, "A relation between distance and radial velocity among extra-galactic nebulae" (1929); see also Hubble's *Realm of the Nebulae* (1936).

Michael Rowan-Robinson in *The Cosmological Distance Ladder* (1985) discusses the many problems involved in measuring extragalactic distances and the resulting uncertainty in the value of the Hubble term.

7. George Gamow's calculations in "The evolution of the universe" (1948), revised by his colleagues Ralph Alpher and Robert Herman in "Evolution of the universe" (1948), predicted a present temperature of the cosmic background radiation of 5 degrees above absolute zero. The hot big bang was described in Gamow's popular book *The Creation of the Universe* (1952). Steven Weinberg's *First Three Minutes: A Modern View of the Origin of the Universe* (1977) presents a nontechnical, more recent treatment of the early universe. Arno Penzias and Robert Wilson in "A measurement of excess antenna temperature at 4090 MHz" (1965) reported the detection of the low-temperature background radiation, and more recent observations confirm that this radiation has a temperature of 3 degrees above absolute zero.

8. Hermann Bondi and Thomas Gold published "The steady-state theory of the expanding universe" in 1948 and in the same year Fred Hoyle published "A new model for the expanding universe." The original idea was proposed by Gold. See "Steady state origins: comments I" by Bondi and "Steady state origins: comments II" by Gold in *Cosmology and Astrophysics: Essays in Honor of Thomas Gold,* edited by Yervant Terzian and Elizabeth Bilson. A static universe neither expands nor collapses; a steady-state universe remains unchanged in appearance, though it may expand, collapse, or remain static.

9. MacMillan's cosmological model is discussed by Richard Schlegel in "Steady-state theory at Chicago" (1958).

17. The Cosmic Redshift

1. Cosmic redshift is best interpreted as the result of the expansion of space. Unlike the Doppler redshift, it is not the result of relative motion in space. Distances between comoving galaxies in an expanding homogeneous universe are proportional to the scaling factor R. If R doubles, all extragalactic distances double. The wavelength λ of a ray of light traveling in extragalactic space varies as R. If λ_0 is the present wavelength and R_0 the present value of the scaling factor, then $\lambda/\lambda_0 = R/R_0$. Redshift is defined by $z = (\lambda_0 - \lambda)/\lambda$ and hence the cosmic redshift is $z = R_0/R - 1$. Edwin Hubble's paper "A relation between distance and radial velocity among extra-galactic nebulae" (1929) confirmed theoretical predictions. In the same volume of the *Proceedings of the National Academy of Sciences,* Fritz Zwicky, in a paper "On the red shift of spectral lines through interstellar space" (1929), proposed the first of the tired-light theories to explain the cosmic redshift. These theories attempt to show by various mechanisms how light loses energy without being scattered while traveling in intergalactic space. From the viewpoint of the riddle of darkness, we might classify the tired-light theories with Chéseaux's absorption solution.

2. Hermann Bondi, *Cosmology* (1952), p. 23. The treatment of the riddle in the second edition of 1960 is unchanged. At a meeting of the British Association

at Oxford in 1954 Bondi discussed the riddle of darkness in "Theories of cosmology," and said, "In many ways Olbers' argument is the basis of all modern cosmology and it will accordingly be presented fully . . . It is known that light from a receding source appears to be redder and weaker than if the source were static, that these effects become large as the velocity of the source approaches the velocity of light, and that light from an approaching source is similarly bluer and stronger. If the universe is an expanding system, then the stars of the distant shells (in Olbers' argument) are receding from us, and the far distant shells are receding very fast. The light from these shells will therefore be considerably weaker than according to Olbers' argument, and the intensity of the background light of the sky will be less than calculated before. Accordingly, if the rate of expansion is sufficiently high, the background light of the sky will be as faint as it actually is, and so Olbers' paradox will be resolved. If the universe were contracting, then the light from distant shells would be enhanced and the paradox would be made worse." Bondi at the time was unaware of the earlier work by Halley and Chéseaux. Fred Hoyle in his popular *Frontiers of Astronomy* (1955), p. 304, wrote, "Let us start with an everyday question, one so trivial that probably few have bothered to ask it, and yet one that has the most profound connections with the distant parts of the Universe. This is a question first asked by Olbers in 1826, and recently revived by H. Bondi. Why is the sky dark at night?" Following Bondi's original treatment, numerous discussions such as that by Hoyle have appeared in the literature.

3. In different publications Bondi analyzed Olbers's assumptions in various ways. The list quoted in the text comes from his essay "Astronomy and cosmology" (1955). In a footnote Bondi points out that assumption (1) is the cosmological principle, and (1) and (2) combine to form the perfect cosmological principle.

4. The steady-state universe has the same metric as the de Sitter universe. The scaling factor R is proportional to $\exp(Ht)$, and the Hubble term $H = \dot{R}/R$ has constant value. The horizon of the visible universe in this case lies at the Hubble distance c/H, where recession equals the velocity of light. The observer's backward lightcone stretches out and back in time and has a radial distance

$$r = (1 - e^{Ht})c/H$$

at any instant $t < 0_0$. Thus the lightcone asymptotically approaches the horizon of the visible universe at the Hubble distance c/H. The redshift of a source is $z = e^{-Ht} - 1$, and its distance is hence

$$r = \frac{c}{H}\left(\frac{z}{1 + z}\right)$$

Beyond the horizon lie events that remain forever unobservable by the observer. Most of the stars covering the sky, as seen by the observer, lie close to the horizon, and their number is given by $4\pi(c/H)^2/\sigma$, which has the approximate value 10^{35}, where σ denotes the cross-section of the average star.

5. The integral equations were examined by Gerald Whitrow and B. D. Yal-

lop in "The background radiation in homogeneous isotropic world-models," papers I and II. These authors give references to earlier work. The differential equations governing radiation in a cavity of varying volume, with cosmological applications, were set up in Harrison, "Olbers' paradox" (1964) and "Olbers' paradox and the background radiation in an isotropic homogeneous universe" (1965). In the first paper I wrote: "We now imagine that during the entire history of the universe a single cell of [comoving] volume V is enclosed with walls which are perfectly reflecting from both inside and outside. At any instant an observer outside V finds that the radiation conditions are exactly the same as those found by an observer inside V. The observer outside V tends to evaluate the radiation density by making a time-retarded integration of the Doppler-shifted contributions from distant stars. The observer inside V, however, tends to use the differential equations of classical thermodynamics for evaluating the density of the radiation emitted by the local stars in a cavity of varying volume. Since both methods must yield identical results, I shall adopt the latter method because of its greater simplicity." William Davidson in a letter "Local thermodynamics and the universe" published in *Nature* (1965) pointed out that he had mentioned the possibility of using differential equations in 1962. The integral and differential equation treatments are contrasted and shown formally to be equivalent in my review "Radiation in homogeneous and isotropic models of the universe" (1977). The equivalence of the two methods helps to explain why the curvature of homogeneous space plays no role in determining the density of radiation in space.

Let u denote the radiation density in space, and u^* the radiation density at the surface of a typical star; then the differential equation showing how u changes with time is

$$\frac{d(uR^4)}{dt} = \frac{R^4}{\tau}(u^* - u)$$

where τ is the "fill up time" equal to the background limit divided by the speed of light. As an illustration of the conceptual utility of the differential equation approach, consider the steady-state universe, in which u and u^* are constant, and R varies as $\exp(Ht)$. We find immediately:

$$u = u^*/(1 + 4H\tau)$$

When $4H\tau$ is large compared with unity (the background limit then greatly exceeds the Hubble distance), u is very much less than u^* and the sky is dark at night; when, however, $4H\tau$ is small compared wtih unity (the background limit is then less than the Hubble distance), u almost equals u^*. Clearly, we can design steady-state universes that are either bright or dark with starlight.

18. Energy in the Universe

1. "$E = Mc^2$" in *Science Illustrated*, April 1946, reproduced in *Out of My Later Years* (1950) by Albert Einstein. Note that mass M includes all forms of energy,

and c is the speed of light. Let M_0 be the mass of a body at rest relative to an observer, and M be the mass of the same body in motion at velocity v. The difference $(M - M_0)c^2$ is the kinetic energy, where $M = M_0/(1 - v^2/c^2)^{1/2}$, and M_0 is the rest mass. When v/c is small,

$$\text{kinetic energy} = M_0 c^2 (1 - v^2/c^2)^{-1/2} - 1 = \tfrac{1}{2} M_0 v^2$$

2. Let ρ denote the average density of matter in the universe and a the radiation density constant. The radiation temperature T is given by the relation $aT^4 = \rho c^2$. If for the density ρ we take a value of one hydrogen atom per cubic meter, this relation yields a temperature $T = 20K$.

3. See Harrison, "Olbers' paradox" (1964). This letter to *Nature* discusses the energy argument for the first time and outlines the differential-equation method of studying the properties of radiation in an expanding universe. It concludes with the words, "Thus the radiation level is low in our universe at present and the night sky is dark simply because the stars are so widely separated from each other." The equivalent relations are interesting: the background limit is 10^{23} light-years, whereas the visible universe has a radius 10^{10} light-years, making the background limit 10^{13} times more distant than the horizon of the visible universe; the time to fill the universe with starlight in equilibrium with stellar surfaces is 10^{13} times a luminous lifetime of a star assumed to be 10^{10} years; and finally the universe requires 10^{13} times more stars than at present to cover the entire sky.

4. If R increases as t^n, the Hubble term is $H = n/t$, the deceleration term is $q = -\ddot{R}R/\dot{R}^2 = (1 - n)/n$. Consider two boxes, each of volume V, each containing an identical star of luminosity L that shines for time t. If the first box is static, the total radiation energy in volume V is Lt. If the second box expands from a very small volume in time t, the radiation energy in V is $Lt/(1 + n)$, and is weaker. The boxes of equal volume contain equal numbers of photons emitted in the same way in the past. The only difference between the radiation in the two boxes is the redshift of the photons in the box that has expanded, which reduces the radiation density by a factor $1/(1 + n)$. Generally, we require $H > 0$ and $q > 0$, and therefore $0 < n < 1$. The maximum reduction of the radiation density by expansion is therefore $\tfrac{1}{2}$. An acceptable value of $n = 2/3$, as in the model proposed by Einstein and deSitter, yields a reduction factor 0.6.

Appendix 1. Digges on the Infinity of the Universe

1. This follows the reproduction by Francis R. Johnson and Sandford V. Larkey in "Thomas Digges, the Copernican system, and the idea of the infinity of the universe in 1576," *The Huntington Library Bulletin* (1934): 83–95. Johnson and Larkey write: "This treatise on the Copernican system of the universe was clearly intended by Digges as a sort of stop-gap until he could publish a more important work he was writing. He explains at the beginning of his preface that, while preparing this new edition of his father's *Prognostication*, he noticed its plan of the universe was according to the old Ptolemaic system, and he

had been unwilling to let that be reprinted without inserting also a diagram and description of the universe according to the new Copernican theory, 'to the ende such noble English minds (as delight to reache above the baser sort of men) might not be altogether defrauded of so noble a part of Philosophy.'"

Digges used astronomical symbols to denote the planets, Sun, and Moon; Johnson and Sandford replace these symbols with proper names, and insert an asterisk to show that a change has been made; I have omitted these asterisks.

2. In the omitted two pages Digges compares the Ptolemaic and Copernican systems.

3. *According to whose measures may the deities be set in motion and the orbs receive the laws and preserve the prescribed agreements.*

4. *We set out from the harbor, lands and cities recede from view.* I am indebted to Professor Rex Wallace for translations in notes 3 and 4.

Appendix 2. Halley on the Infinity of the Sphere of Stars

1. In "Dark skies and fixed stars" (1973) Michael Hoskin points out that the *Journal Book* of the Royal Society records that Halley read his two papers on March 9 and 16, 1721.

2. These comments by Halley recall the discussion in the Newton–Bentley correspondence on the gravitational equilibrium and stability of the universe. Newton realized that the assumed state of equilibrium of an unbounded distribution of self-gravitating matter is unstable. Almost all theoreticians until the early decades in this century, including James Jeans, who made a study of this subject, agreed with Newton. Halley alone incorrectly supposed that the equilibrium state of an unbounded self-gravitating system is stable. The theory of general relativity has shown that an unbounded, uniform, self-gravitating system has in fact no equilibrium state, either stable or unstable. An equilibrium state can be contrived by means of a "cosmological term," as in the Einstein static universe, and this state is unstable.

3. "To biquadrate," a term in use since 1694, means to raise to the fourth power. Halley's second paper shows more clearly that he imagined the stars occupying shells constructed from spherical surfaces, each shell having a thickness equal to the separating distance between neighboring stars. Now, the number of stars occupying a shell is proportional to its surface area and hence proportional to the square of the radius of the shell, and therefore it is not clear why Halley thought the number of contained stars must be proportional to the fourth power.

4. Halley knew that the apparent disk-area of an individual star and the light received from it diminish in proportion to the inverse square of distance. He also knew that the *apparent* separating distance between neighboring stars diminishes as the inverse of distance. The apparent disk-area thus decreases faster than the apparent separating distance, and he incorrectly concluded that this simple geometric fact explains why the gaps between stars remain unfilled and why the heavens are dark. It seems in this confused argument that he failed

to realize that in each shell we must compare the apparent disk-area of the stars with their apparent separating area. Because the number of stars in a shell is proportional to the square of the shell's distance, and because the light received from a star, like the apparent disk-area of a star, is proportional to the inverse square of distance, we see that all shells contribute equal amounts of light and star-coverage.

5. Possibly Halley realized that his geometrical argument might fail to convince. This may explain why he fell back on the intuitive argument previously used by Thomas Digges, stating that the light from distant stars is "not sufficient to move our Sense." Like Digges, but with less excuse, he overlooked the accumulative effect of many small sources.

6. The immediate neighbors in two-dimensional and three-dimensional uniform arrays cannot have identical separations. Newton considered this problem in the case of stars after corresponding with Bentley in 1692–1693 (see Michael Hoskin, "Newton, providence and the universe of stars," 1977). It is difficult to understand why a practical man like Halley bothered with so profitless a geometrical investigation. An analogous situation exists when grains of sand are scattered more or less uniformly on the floor; in this case we may discuss the average separating distance of grains and also their number per unit area, and remain quite unconcerned with the geometrical problem of why all neighboring grains cannot have identical separations. A few lines later Halley becomes more practical and comments on the fact that the brightest stars have unequal spacings in the sky.

7. John Herschel confirmed that stars of the first magnitude are 100 times brighter than those of the sixth magnitude.

8. Halley's second paper clarifies certain parts of his first paper, which may be the reason why he wrote it. But it fails to provide an acceptable analysis of the riddle, and to the attentive reader the proposed solution remains as obscure as in the first paper.

Appendix 3. Chéseaux Explains the Riddle of Darkness

1. In this rendition into English I have used parts of Debbie Van Dam's helpful unpublished translation. Chéseaux uses the term "force" for either the intensity or the flux. We note that intensity (flux per unit solid angle) as defined nowadays does not vary with distance from a point source, whereas the flux F (energy incident per unit time per unit area) varies as the inverse of the square of distance. If $I_\odot$ denotes the intensity of the Sun's radiation, the flux at distance q is $F_\odot = \pi I_\odot a^2/q^2$, where a is the radius of the Sun. Very slight alterations to Chéseaux's terminology have been made to avoid conflict with modern definitions.

2. Let $L_\odot$ be the luminosity of the Sun. The "quantity of light" (or luminous flux) from the Sun is $F_\odot = L_\odot/4\pi q_\odot^2$, where $q_\odot$ represents the distance of the Sun from the Earth. The flux from a star of similar luminosity at distance q is $L_\odot/4\pi q^2$, or $F_\odot(q_\odot/q)^2$, as stated by Chéseaux.

3. The Cartesian planetary vortices filled the whole of interstellar space, and the diameter of a vortex was therefore a measure of the separating distance between stars. Chéseaux's shells have a thickness equal to the average separating distance between stars and each consists of a single layer of planetary vortices.

4. Let $d = 2a$ be the actual diameter of a star. A star at distance q has an apparent or angular diameter d/q. Chéseaux assumes without comment that stars are uniformly distributed; hence the number contained in a shell is proportional to q^2, as he states. The luminous flux from a layer at distance q is proportional to the number of stars contained in that layer multiplied by the square of the apparent diameter $(d/q)^2$ of each star. The flux from a layer is hence proportional to q^2 multiplied by $(d/q)^2$, and is therefore independent of the value of q.

5. Chéseaux states that the luminous flux from a layer equals the number of contained stars multiplied by $F_{\odot}(q_{\odot}/q)^2$. The nearest layer contains approximately 10 first magnitude stars at a distance 200,000 times the distance of the Sun, and the flux from this layer is $\frac{1}{4} \times 10^{-9}$ times the flux $F_{\odot}$ from the Sun. The solid angle of the celestial sphere is roughly 1.8×10^5 times that of the Sun, and Chéseaux argues that the number of shells is therefore $1.8 \times 10^5 \times 4 \times 10^9 = 7.2 \times 10^{14}$.

6. If light is diminished by one thirty-third each time it passes through a layer, then in effect almost no light comes to us from layers farther away than 33 times the distance of the nearest stars. See note 1, Chapter 8.

7. Angular notation in the sexagesimal system is degrees, minutes, seconds, 1/60th seconds, 1/3600th seconds.

8. Bouguer's published work of 1729 is translated into English by W. E. Knowles Middleton in *Pierre Bouguer's Optical Treatise on the Gradation of Light* (University of Toronto Press, Toronto, 1961). See also Stanley Jaki's informative chapter 5 in *The Paradox of Olbers' Paradox* (1969).

Appendix 4. Olbers Revives the Riddle of Darkness

1. J. E. Bode, editor of *Astronomisches Jahrbuch*, published Olbers's paper in the 51st issue, 1823, containing advance astronomical information for 1826. In the following rendition I have taken note of the translation in the *Edinburgh New Philosophical Journal* (1826) and have used with slight alterations several passages in Deborah Schneider's helpful unpublished translation.

2. In this quotation by Olbers I have taken note, but not followed exactly, W. Hastie's translation of Kant's *Universal Natural History and Theory of the Heavens* in his *Kant's Cosmogony* (1900), pp. 138–140.

3. In this impressionistic quotation of Halley's remarks (see the opening paragraph of Halley's first 1721 paper), Olbers uses the word *Schwerpunkt* (center of gravity), a term probably unfamiliar to Halley.

4. Olbers uses n to denote the number of first magnitude stars; to avoid confusion I have changed this symbol to N_1. The following remarks make clear

Olbers's treatment. He first constructs shells of equal thickness with imaginary spheres of increasing radius. Let q_1 be the radius of the first sphere with N_1 stars supposed fixed to its surface. Each star on this sphere of area $4\pi q_1^2$ presents a disk-area πa^2, and thus subtends a fraction $\pi a^2/4\pi q_1^2 = a^2/4q_1^2$ of the celestial sphere (the whole sky). The total fraction for N_1 stars is hence $N_1 a^2/4q_1^2$. Olbers calls the quantity $\delta = a/q_1$ the radius of a star and then confuses the discussion by referring to it as diameter. He measures distances in terms of q_1, setting q_1 (the distance of Sirius) equal to unity. The second sphere of radius q_2 (equal to $2q_1$) has N_2 stars fixed to its surface. The fraction of the sky covered by all stars on this sphere is $N_2 a^2/4q_2^2$; for uniformly distributed stars we have $N_2 = N_1 q_2^2/q_1^2$, thus yielding $N_1 a^2/q_1^2$. This result is the same as that obtained for the first sphere. We see that all spheres contribute equally, and therefore the fraction of the sky covered by the stars on m spheres, reaching out to distance $q_m = mq_1$, is

$$mN_1 a^2/4q_1^2$$

This fraction equals unity when $m = 4q_1^2/N_1 a^2$, and hence the background limit (not calculated by Olbers) is $\lambda = mq_1$, or

$$\lambda = 4q_1^3/N_1 a^2.$$

Using for N_1 a value of order 10 and for q_1 (the average separating distance between stars) a value of order 5 light-years, we find that the number of shells needed to cover the sky is 2×10^{15}, and the background limit is 10^{16} light-years. Olbers, unlike Chéseaux, does not calculate these results.

5. Intersecting light rays cannot interfere with one another and scatter in the manner considered by Olbers. Rays of light entering the Gregorian telescope cross one another before being brought to a focus in the eye or on a photographic plate, whereas in the Cassegrain telescope they do not cross at an intermediate focus. The papers by Henry Kater, referred to by Olbers, claiming to have detected the effect of rays scattering one another, can be found in the *Philosophical Transactions* 103 (1813):206; *ibid.* 184 (1814): 231. Olbers appears not to recognize that even should rays scatter one another in the manner supposed, radiation would not be lost but be merely redirected, and the sky would still blaze at every point with starlight.

6. On pages 117 to 119 of his paper Olbers gives an unenlightening and arithmetically inaccurate treatment of radiative transfer, taking into account the effect of interstellar absorption. The following gives a clearer analysis. Let L_0 be the luminosity of a star (this is the energy radiated each second from its surface) and let L by this radiant energy when it has traveled distance q. Olbers uses an equation of the form

$$dy = -ydq/\mu$$

where $y = L/L_0$ and μ is the mean free path for absorption. He employs a volume absorption coefficient $a = 1/\mu$, and guesses that of every 800 rays emitted from Sirius, 1 ray is lost in traversing a distance equal to that of Sirius from the Sun. In effect he adopts an absorption limit or cut-off distance equal to 800 times the distance of Sirius, or 7,000 light-years, with Sirius at 8.6 light-years. Chéseaux assumed a much larger absorption coefficient, and his absorption limit is roughly only 33 times the distance to Sirius. When all distances are normalized to that of Sirius, Olbers's absorption coefficient is $a = 1/800 = 0.00125$, from $dy/y = -1/800$. Olbers uses a tedious procedure and finds, working to 10 places of decimals, an arithmetically incorrect value for a. In his analysis he fails to show that the star-covered sky is dark because the absorption limit is much smaller than the background limit. This is discussed in the notes to Chapter 8.

7. The conclusion that stars at 30,000 times the distance of Sirius have a brightness equal to that of the Moon is of doubtful significance. Unlike Chéseaux, Olbers apparently fails to appreciate that all stars beyond the absorption limit contribute little to the light of the night sky. Using his value for the absorption coefficient, we note that all stars beyond 800, 1,600, 2,400, 3,200, and 4,000 times the distance of Sirius contribute 37, 14, 5, 2, and 0.7 percent respectively to the total amount of starlight. The stars beyond 30,000 times the distance of Sirius contribute only 5×10^{-17} of the total.

8. In this passage, which I have omitted, Olbers presents on page 120 evidence to show that the empty gaps between stars are never completely dark, even on the clearest moonless nights, because of the atmospheric scattering of starlight.

Appendix 5. Kelvin on an Old and Celebrated Hypothesis

1. This result neglects the geometric occultation of stars. The neglect may easily be remedied. The fraction of the sky covered by stars in a shell of radius q must be multiplied by the probability $\exp(-q/\lambda)$ that these stars are occulted by the stars occupying a sphere of radius q. In this expression $\lambda = 1/\pi na^2$ is the mean free path of a light ray. Instead of Kelvin's equation (10) we obtain

$$\alpha = 1 - e^{-r/\lambda} \qquad (A)$$

for the fraction of the sky covered by stars out to distance r, as shown in note 5, Chapter 14. When the ratio r/λ is small, as assumed by Kelvin in his treatment, then

$$\alpha = r/\lambda = 3Na^2/4r^2$$

2. Kelvin recognized that the fraction of the sky covered by stars equals the brightness ratio of the starlit night sky and the Sun's disk. This may be demonstrated formally as shown in note 5 of Chapter 14. The radiation density u of starlight from stars out to distance r is found to be

$$u = u^*(1 - e^{-r/\lambda}) \qquad (B)$$

where $u^* = L/\pi a^2 c$ is the radiation density at the surface of a star. Hence $u = u^*$ in a distribution of stars extending beyond the background limit λ. From equations (A) and (B) we obtain

$$\alpha = u/u^* \tag{C}$$

as recognized by Kelvin.

3. Because the aim is to compare the apparent brightness of the starlit sky with the brightness of the Sun's disk, Kelvin considered the eclipsing of geometric disks and correctly ignored the eclipsing of diffraction disks. His remarks on this point, however, are not very clear.

4. Relativity theory had yet to stress the universality of $E = Mc^2$, thus providing a true maximum luminous lifetime $M_\odot c^2/L_\odot$ of order 10^{13} years, where $M_\odot$ is the mass and $L_\odot$ the luminosity of the Sun. Given even this extreme value of the luminous lifetime, Kelvin would have arrived at essentially the same conclusion.

Bibliography

Abbott, E. A. *Flatland*. New York: Dover, 1952.

Abell, G. O. *Exploration of the Universe*. New York: Holt, Rinehart and Winston, 1975.

Aiton, E. J. *The Vortex Theory of Planetary Motions*. London: Macdonald, 1972.

Alexander, H. G., ed. *The Leibniz–Clarke Correspondence, Together with Extracts from Newton's "Principia" and "Opticks."* Manchester: Manchester University, 1956.

Allerton, M. *Origin of Poe's Critical Theory*. New York: Russell and Russell, 1925.

Alpher, R. A., H. Bethe, and G. Gamow. "The origin of the chemical elements." *Physical Review* 73 (1948): 803.

Alpher, R. A., and R. C. Herman. "Evolution of the universe." *Nature* 162 (1948): 774.

Ames, J. S., ed. *Prismatic and Diffraction Spectra: Memoirs by Joseph von Fraunhofer*. New York: Harper, 1898.

Andrade, E. N. da C. "Newton." In *Newton Tercentenary Proceedings of the Royal Society*, p. 3. Cambridge: Cambridge University Press, 1947.

———— "Robert Hooke." *Proceedings of the Royal Society* 201A (1950): 439.

———— "Doppler and the Doppler effect." *Endeavour*, January 1959, p. 14.

Ariew, R. *Uneasy Genius: The Life and Work of Pierre Duhem*. Dordrecht, Holland: Nijhoff, 1984.

Aristotle. *On the Heavens (De Caelo)*. Trans. W. K. C. Guthrie. Cambridge, Mass.: Harvard University Press, 1939.

———— *On the Soul*. Trans. J. A. Smith. Oxford: Clarendon Press, 1908.

———— *Sense and Sensible*. Trans. J. E. Beare. Oxford: Clarendon Press, 1908.

Armitage, A. *A Century of Astronomy*. London: Samson Low, 1950.

———— *Copernicus: The Founder of Modern Astronomy*. New York: Thomas Yoseloff, 1957.

———— *John Kepler*. London: Faber and Faber, 1966.

———— *Edmond Halley*. London: Thomas Nelson, 1966.

———— *William Herschel*. London: Thomas Nelson, 1962.

Arrhenius, S. "Infinity of the universe." *The Monist* 21 (1911): 161.

Ashbrook, J. *The Astronomical Scrapbook: Skywatchers, Pioneers, and Seekers in Astronomy*. New York: Cambridge University Press, 1980.

Austin, W. H. "Isaac Newton on science and religion." *Journal of the History of Ideas* 31 (1970): 521.

Baade, W. "A revision of the extra-galactic distance scale." *Transactions of the International Astronomical Union* 8 (1952): 397.

―――― *Evolution of Stars and Galaxies*, ed. C. Payne-Gaposchkin. Cambridge, Mass.: Harvard University Press, 1963.

Bailey, C. *The Greek Atomists and Epicurus*. Oxford: Clarendon Press, 1928.

Ball, R. S. *In Starry Realms*. Philadelphia: Lippincott, 1892.

―――― *In the High Heavens*. London: Isbister, 1894.

Barnard, E. E. "On the vacant regions of the Milky Way." *Popular Astronomy* 14 (1906): 579.

Beaver, H., ed. *The Science Fiction of Edgar Allan Poe*. Harmondsworth: Penguin, 1976.

Beer, A., and P. Beer, eds. *Kepler: Four Hundred Years*. Oxford: Pergamon Press, 1975.

Begbie, G. H. *Seeing and the Eye: An Introduction to Vision*. Garden City, N.Y.: Natural History Press, 1969.

Bell, A. E. *Christian Huygens and the Development of Science in the Seventeenth Century*. London: Arnold, 1947.

Bennett, J. A. "Cosmology and the magnetical philosophy, 1640–1680." *Journal for the History of Astronomy* 12 (1981): 165.

Bentley, R. *A Confutation of Atheism from the Origin and Frame of the World*. London, 1693. Reprinted in *Isaac Newton's Papers and Letters on Natural Philosophy*, ed. I. B. Cohen. Cambridge, Mass.: Harvard University Press, 1958.

―――― *The Works of Richard Bentley*. Ed. A. Dyce. 8 vols. New York: AMS Press, 1966.

Benton, R. P., ed. *Poe as Literary Cosmologer. Studies on Eureka: A Symposium*. Hartford, Conn.: Transcendental Books, 1975.

Berendzen, R., R. Hart, and D. Seeley. *Man Discovers the Galaxies*. New York: Science History Publications, 1976.

Bernstein, J. *Einstein*. New York: Viking, 1973.

Berry, A. *A Short History of Astronomy: From Earliest Times through the Nineteenth Century*. 1898. Reprinted, New York: Dover Publications, 1961.

Bok, B. J., and P. F. Bok. *The Milky Way*. Cambridge, Mass.: Harvard University Press, 1957.

Bondi, H. *Cosmology*. Cambridge: Cambridge University Press, 1952; 2nd ed. 1960.

―――― "Theories of cosmology." *Advancement of Science* 12 (1955): 33.

―――― "Astronomy and cosmology." In *What Is Science?* ed. J. R. Newman. New York: Simon and Schuster, 1955.

―――― *The Universe at Large*. Garden City, N.Y.: Doubleday, 1959.

Bondi, H., and T. Gold, "The steady-state theory of the expanding universe." *Monthly Notices of the Royal Astronomical Society* 108 (1948): 252.

Bonnor, W. "On Olbers' Paradox." *Monthly Notices of the Royal Astronomical Society* 128 (1964): 33.

Born, M. *Einstein's Theory of Relativity*. New York: Dover Publications, 1962.

Boyle, R. *Selected Philosophical Papers of Robert Boyle*. Ed. M. A. Stewart. Manchester: Manchester University Press, 1979.

Bradley, J. "A letter to Dr Edmund Halley, giving an account of a new-discovered motion of the fixed stars." *Philosophical Transactions* 35 (1729): 637.

Brewster, D. *Memoirs of the Life, Writings, and Discoveries of Sir Isaac Newton.* 2 vols. Edinburgh: Constable, 1855. Reprinted with introduction by R. S. Westfall, New York: Johnson Reprint Corporation, 1965.

Brickman, R. "On physical space, Francesco Patrizi." *Journal of the History of Ideas* 4 (1943): 224.

Brooks, C. M. "The cosmic God: science and the creative imagination in *Eureka.*" In *Poe as Literary Cosmologer. Studies on Eureka: A Symposium,* ed. R. P. Benton. Hartford, Conn.: Transcendental Books, 1975.

Bruno, G. *Jordani Bruni Nolani Opera Latine Conscripta.* 3 vols. Ed. F. Fiorentino (1879–1891). Stuttgart: Friedrich Frommann Verlag, Gunther Holzboog, 1962.

——— *Gesammelte Werke,* ed. L. Kuhlenbeck. 2nd ed. Jena: Eugen Diederichs, 1904.

Burchfield, J. D. *Lord Kelvin and the Age of the Earth.* New York: Science History Publications, 1975.

Burger, D. *Sphereland: A Fantasy about Curved Spaces and an Expanding Universe.* Trans. C. J. Rheinboldt. New York: Thomas Y. Crowell, 1965.

Burnet, J. *Early Greek Philosophy.* New York: Meridian Books, 1957.

Burtt, E. A. *The Metaphysical Foundations of Modern Physical Science.* London: Routledge and Kegan Paul, 1924.

Butterfield, H. *The Origins of Modern Science 1300–1800.* New York: Macmillan, 1957.

Buttmann, G. *The Shadow of the Telescope: A Biography of John Herschel.* Trans. B. E. J. Pagel. Ed. D. S. Evans. New York: Scribner's Sons, 1970.

Cantor, G. N., and M. J. S. Hodge. *Conceptions of Ether: Studies in the History of Ether Theories 1740–1900.* New York: Cambridge University Press, 1981.

Caspar, M. *Kepler.* Trans. and ed. C. D. Hellman. London: Abelard-Schuman, 1959.

Chambers, G. F. *Descriptive Astronomy.* Oxford: Clarendon Press, 1867.

——— *The Story of the Comets.* Oxford: Oxford University Press, 1909.

Charlier, C. V. L. "Wie eine unendliche Welt aufgebout sein kann." *Arkiv för Matematik, Astronomi och Fysik* 4, no. 24 (1908).

——— "How an infinite world may be built up." *Arkiv för Matematik, Astronomi och Fysik* 16, no. 22 (1922).

——— "On the structure of the universe." *Publications of the Astronomical Society of the Pacific* 37 (1927): 53, 115; 37 (1925): 177.

Chéseaux, J. P. Loys de. *Traité de Comète.* Lausanne: M. M. Bousequet, 1744.

Chincarini, G., and H. J. Rood. "The cosmic tapestry." *Sky and Telescope* 59 (1980): 354.

Clagett, M. *The Science of Mechanics in the Middle Ages.* Madison: University of Wisconsin Press, 1959.

Clayton, D. *The Dark Night Sky: A Personal Adventure in Cosmology.* New York: Quadrangle, 1975.

Clerke, A. M. *The System of the Stars.* London: Longmans, Green, 1890.

—— *A Popular History of Astronomy during the Nineteenth Century.* London: Black, 1893.

—— *The Herschels and Modern Astronomy.* London: Cassell, 1901.

Clerke, A. M., A. Fowler, and J. E. Gore. *Astronomy.* New York: Appleton, 1898.

Cohen, I. B. "Roemer and the first determination of the velocity of light (1676)." *Isis* 31 (1940): 327.

—— "Newton in the light of recent scholarship." *Isis* 51 (1960): 489.

—— "Isaac Newton's *Principia,* the scriptures, and the divine providence." In *Philosophy, Science, and Method,* ed. S. Morgenbesser. New York: Martin's Press, 1969.

—— *The Newtonian Revolution: With Illustrations of the Transformation of Scientific Ideas.* Cambridge: Cambridge University Press, 1980.

Collier, K. B. *Cosmogonies of Our Fathers: Some Theories of the Seventeenth and the Eighteenth Centuries.* New York: Columbia University Press, 1934.

Comte, A. *The Essential Comte,* Ed. S. Andreski. Trans. M. Clarke. London: Croom Held, 1974.

—— *Auguste Comte and Positivism: The Essential Writings.* Ed. G. Lenzer. New York: Harper, 1975.

Connor, F. W. "Poe and John Nichol: notes on a source of *Eureka.*" In *All These to Teach: Essays in Honor of C. A. Robertson,* ed. R. A. Bryan, A. C. Morris, A. A. Murphree, and A. L. Williams. Gainesville: University of Florida Press, 1965.

Copernicus, N. *On the Revolutions.* Ed. J. Dobrzycki. Trans. E. Rosen. Baltimore: Johns Hopkins University Press, 1978.

—— *Revolutions of the Heavenly Spheres.* In *Great Books of the Western World,* vol. 16. Chicago: Encyclopaedia Britannica, 1952.

Cornford, F. M. "The invention of space." In *Essays in Honour of Gilbert Murray,* ed. H. A. L. Fisher. London: Allen and Unwin, 1936, p. 215.

—— *Before and after Socrates.* Cambridge: Cambridge University Press, 1965.

Crombie, A. C. *Robert Grosseteste and the Origins of Experimental Science 1180–1700.* Oxford: Clarendon Press, 1953.

—— *Augustine to Galileo.* Vol. 1, *Science in the Middle Ages.* Vol. 2, *Science in the Later Middle Ages and Early Modern Times.* New York: Doubleday, 1959.

Curtis, H. D. "Dimensions and structure of the Galaxy." *Bulletin of the National Research Council of the National Academy of Sciences* 2 (1921): 194.

Cusa, N. *Of Learned Ignorance.* Trans. G. Heron. New Haven: Yale University Press, 1954.

Dampier, W. C. *A History of Science and Its Relations with Philosophy and Religion.* Cambridge: Cambridge University Press, 1966.

Dante. *The Comedy of Dante Alighieri: Vol. 3, Paradise.* Trans. D. Sayers and B. Reynolds. Harmondsworth: Penguin, 1962.

Davidson, W. "The cosmological implications of the recent counts of radio sources: II. An evolutionary model." *Monthly Notices of the Royal Astronomical Society* 124 (1962): 79.

——— "Local thermodynamics and the universe." *Nature* 206 (1965): 249.

de Santillana, G. *The Crime of Galileo*. Chicago: University of Chicago Press, 1955.

Descartes, R. *Discourse on Method, Optics, Geometry, and Meteorology*. Trans. P. J. Olscamp. Indianapolis: University of Indianapolis, 1965.

——— *The Philosophical Works of Descartes*. 2 vols. Trans. E. S. Haldane and G. R. T. Ross. Cambridge: Cambridge University Press, 1911.

——— *Oeuvres de Descartes*. Ed. C. Adam and P. Tannery. 11 vols. 1897–1913. Revised reprint, Paris: Librairie Philosophique, J. Vrin, 1974.

de Sitter, W. "On Einstein's theory of gravitation and its astronomical consequences." *Monthly Notices of the Royal Astronomical Society* 76 (1916): 49; 77 (1916): 155; 78 (1917): 3.

——— "On the curvature of space." *Proceedings of the Royal Academy of Amsterdam* 20 (1917): 229.

——— *Kosmos*. Cambridge, Mass.: Harvard University Press, 1932.

de Vaucouleurs, G. *The Discovery of the Universe*. New York: Macmillan, 1957.

——— "The case for a hierarchical cosmology." *Science* 167 (1970): 1203.

DeWitt, N. W. *Epicurus and His Philosophy*. Minneapolis: University of Minnesota Press, 1954.

Dick, S. J. *Plurality of Worlds: The Origin of the Extraterrestrial Life Debate from Democritus to Kant*. New York: Cambridge University Press, 1982.

Dick, T. *Celestial Scenery; Or the Wonders of the Planetary System Displayed; Illustrating the Perfections of Deity and a Plurality of Worlds*. Brookfield, Mass.: Merriam, 1838.

——— *The Sidereal Heavens and Other Subjects Connected with Astronomy*. New York: Harper, 1840.

——— *The Complete Works of Thomas Dick*. 2 vols. Cincinnati: Applegate, 1850.

Digges, T. "A perfit description of the caelestiall orbes." In *Prognostication Everlastinge*. London, 1576. Reprinted in F. R. Johnson and S. V. Larkey, "Thomas Digges, the Copernican system, and the idea of the infinity of the universe in 1756." *Huntington Library Bulletin*, no. 5 (April 1934): 69–117.

Dijksterhuis, E. J. *The Mechanization of the World Picture*. Trans. C. Dikshoorn. Oxford: Clarendon Press, 1961.

Dobbs, B. J. T. *The Foundations of Newton's Alchemy: The Hunting of the Greene Lyon*. New York: Cambridge University Press, 1975.

Donne, J. *The Poems of John Donne*. Ed. H. H. C. Grierson. 2 vols. Oxford: Clarendon Press, 1912.

Doppler, C. "Ueber das farbige Licht des Dopplesterne und einige anderer Gestirne des Himmels" (On the colored light of the double stars and the celestial constellations). *Abhandlungen der Königlichen Böhmischen Gesellschaft der Wissenschaften* (*Proceedings of the Royal Bohemian Society of Learning*) 2 (1843): 465.

Doré, G. *The Doré Illustrations for Dante's Divine Comedy.* New York: Dover Publications, 1976.

Douglas, A. V. *The Life of Arthur Stanley Eddingtion.* London: Thomas Nelson, 1956.

Drake, S. *Discoveries and Opinions of Galileo.* Garden City, N.Y.: Doubleday, 1957.

—— *Galileo at Work: His Scientific Biography.* Chicago: University of Chicago Press, 1978.

Dreyer, J. L. E. "On the multiple tail of the great comet of 1744." *Copernicus, An International Journal of Astronomy* 3 (1884): 104.

—— *History of the Planetary Systems from Thales to Kepler.* Cambridge: Cambridge University Press, 1906. Revised by W. H. Stahl, *A History of Astronomy from Thales to Kepler.* New York: Dover Publications, 1953.

Duhem, P. *Le Système du Monde: Histoire des Doctrines Cosmologiques de Platon à Copernic.* 10 vols. Paris: Librairie Scientifique Hermann, 1913–1959.

—— *To Save the Phenomena: An Essay on the Idea of Physical Theory from Plato to Galileo.* Trans. E. Doland and C. Maschler. Chicago: Chicago University Press, 1969.

—— *The Evolution of Mechanics.* Trans. M. Cole. Alphen aan den Rijn, Netherlands, 1980.

—— *Medieval Cosmology: Theories of Infinity, Place, Time, Void, and the Plurality of Worlds.* Ed. and trans. R. Ariew. Chicago: University of Chicago Press, 1985.

Dunkin, E. *The Midnight Sky: Familiar Notes on the Stars and Planets.* London: Religious Tract Society, 1891.

Dyce, A. *The Works of Richard Bentley.* Vol. 3, *Sermons Preached at Boyle's Lectures.* London: Macpherson, 1838.

Eddington, A. S. *Space, Time, and Gravitation: An Outline of the General Relativity Theory.* Cambridge: Cambridge University Press, 1920.

—— *The Mathematical Theory of Relativity.* Cambridge: Cambridge University Press, 1923.

—— *The Internal Constitution of the Stars.* Cambridge: Cambridge University Press, 1926. Reprinted, New York: Dover Publications, 1959.

—— *The Expanding Universe.* Cambridge: Cambridge University Press, 1933.

Einstein, A. "Cosmological considerations on the general theory of relativity." 1917. Reprinted in H. A. Lorentz, A. Einstein, H. Minkowski, and H. Weyl, *The Principle of Relativity: A Collection of Original Memoirs on the Special and General Theory of Relativity,* trans. W. Perrett and G. B. Jeffrey. New York: Dover Publications, 1952.

—— *Sidelights on Relativity.* New York: Dutton, 1923.

—— *The World as I See It.* New York: Crown, 1934.

—— "E = Mc²." In *Out of My Later Years.* New York: Philosophical Library, 1950.

—— *The Meaning of Relativity.* Princeton: Princeton University Press, 1953.

———— *Relativity: The Special and General Theory.* Trans. R. W. Lawson. London: Methuen, 1955.

Einstein, A., and W. de Sitter, "On the relation between the expansion and the mean density of the universe." *Proceedings of the National Academy of Sciences* 18 (1932): 213.

Enriques, F., and G. de Santillana. *Storia del Pensiero Scientifico.* Vol. 1, *Il Mono Antico.* Milan: Treves-Treccani-Tumminelli, 1932.

Epicurus. *Epicurus: The Extant Remains.* Trans. C. Bailey. Oxford: Clarendon Press, 1926.

Fellows, O. E., and S. F. Milliken. *Buffon.* New York: Twayne, 1972.

Figala, K. "Newton as alchemist." *Journal of the History of Science* 15 (1977): 192.

Fontenelle, B. de. *A Plurality of Worlds.* Trans. J. Glanville. 1688. Reprinted, London: Nonsuch Press, 1929.

———— *Entretiens sur la Pluralité des Mondes.* Ed. R. Shackleton. Oxford: Clarendon Press, 1955.

Forbes, G. "Molecular dynamics." *Nature* 32 (1885): 461, 508, 601.

Force, J. E. *William Whiston, Honest Newtonian.* New York: Cambridge University Press, 1985.

Fournier d'Albe, E. E. "The infra-world." *English Mechanic and World of Science,* September 14, 1906, p. 127; September 21, p. 153; September 28, p. 177; October 5, p. 201; October 12, p. 225; October 19, p. 249; October 26, p. 271.

———— "The supra-world." *English Mechanic and World of Science,* March 29, 1907, p. 178; April 5, p. 202; April 12, p. 224; April 19, p. 248; April 26, p. 272; May 3, p. 293.

———— *Two New Worlds.* London: Longmans, Green, 1907.

Frank, P. *Einstein: His Life and Times.* New York: Knopf, 1947.

Frankland, W. B. "Notes on the parallel axiom." *Mathematical Gazette* 7 (1913): 136.

Friedmann, A. "On the curvature of space." Trans. B. Doyle. In *A Source Book in Astronomy and Astrophysics, 1900–1975,* eds. K. R. Lang and O. Gingerich. Cambridge, Mass.: Harvard University Press, 1979.

Gade, J. A. *The Life and Times of Tycho Brahe.* Princeton: Princeton University Press, 1947.

Gale, R. M., ed. *The Philosophy of Time: A Collection of Essays.* New Jersey: Humanities Press, 1968.

Galileo Galilei. *Le Opera di Galileo Galilei.* 20 vols. Florence: Edizione Nazionale, 1890–1909.

———— *The Starry Messenger.* In *Discoveries and Opinions of Galileo,* trans. S. Drake. Garden City, N.Y.: Doubleday, 1957.

———— *Dialogue Concerning the Two Chief World Systems—Ptolemaic and Copernican.* Trans. S. Drake. Foreword A. Einstein. Berkeley: University of California Press, 1953.

———— *Dialogue Concerning Two New Sciences.* Trans. H. Crew and A. de Salvio. New York: Dover Publications, 1954.

Gamow, G. "The evolution of the universe." *Nature* 162 (1948): 680.

———— *The Creation of the Universe.* New York: Macmillan, 1952.

Gershenson, D. E., and D. A. Greenberg. *Anaxagoras and the Birth of Physics.* New York: Blaisdell, 1964.

Gilbert, W. *On the Magnet, Magnetick Bodies Also, and on the Great Magnet Earth; A New Physiology, Demonstrated by Many Arguments and Experiments.* Trans. S. P. Thompson. London: Chiswick Press, 1900. Reprinted, New York: Basic Books, 1958.

Gillispie, C. C. "Fontenelle and Newton." In *Isaac Newton's Papers and Letters on Natural Philosophy,* ed. I. B. Cohen and R. E. Schofield. Cambridge, Mass.: Harvard University Press, 1958.

Gimpel, J. *The Medieval Machine: The Industrial Revolution of the Middle Ages.* New York: Holt, Rinehart and Winston, 1976.

Gingerich, O. "Charles Messier and his catalog." *Sky and Telescope* 12 (1953): 255, 288.

———— "Johannes Kepler and the New Astronomy." *Quarterly Journal of the Royal Astronomical Society* 13 (1972): 345.

———— "Astronomy three hundred years ago." *Nature* 255 (1975): 602.

Goldman, M. *The Demon in the Aether: The Story of James Clerk Maxwell.* Edinburgh: Harris, 1983.

Gore, J. E. *Planetary and Stellar Studies.* London: Roper and Drowley, 1888.

———— *The Scenery of the Heavens: A Popular Account of Astronomical Wonders.* London: Roper and Drowley, 1890.

———— *The Visible Universe: Chapters on the Origin and Construction of the Heavens.* New York: Macmillan, 1893.

———— *Studies in Astronomy.* London: Chatto and Windus, 1904.

Grant, E. "Late medieval thought, Copernicus, and the scientific revolution." *Journal of the History of Ideas* 23 (1962): 197.

———— "Medieval and seventeenth century conceptions of an infinite void space beyond the cosmos." *Isis* 60 (1969): 39.

———— *Physical Science in the Middle Ages.* New York: Wiley, 1971.

————, ed. *A Source Book in Medieval Science.* Cambridge, Mass.: Harvard University Press, 1974.

———— *Much Ado about Nothing: Theories of Space and Vacuum from the Middle Ages to the Scientific Revolution.* New York: Cambridge University Press, 1981.

Grant, R. *History of Physical Astronomy: From the Earliest Ages to the Middle of the Nineteenth Century.* London: Bohn, 1852.

Gregory, R. L. *Eye and Brain: The Psychology of Seeing.* New York: McGraw-Hill, 1966.

Gregory, R. L., and E. H. Gombrich, eds. *Illusion in Nature and in Art.* New York: Scribner's Sons, 1973.

Guericke, O. von (Ottonis de Guericke). *Experimenta Nova (ut Vocantur) Magdeburgica de Vacuo Spatio.* Amsterdam, 1672. Reprinted, Aalen: Otto Zellers Verlagsbuchhandlung, 1962.

Guerlac, H., and M. C. Jacob, "Bentley, Newton, and providence (the Boyle Lectures once more)." *Journal for the History of Ideas* 30 (1969): 307.

Guillemin, A. *The World of Comets.* Ed. and trans. J. Glaisher. London: Samson Low, 1877.

Gunther, R. T. *Early Science at Oxford.* Vol. 8, *The Cutler Lectures of Robert Hooke.* Oxford: Clarendon Press, 1931.

Haber, F. C. *The Age of the World: Moses to Darwin.* Baltimore: Johns Hopkins Press, 1959.

Hahm, D. E. *The Origins of Stoic Cosmology.* Columbus, Ohio: Ohio University Press, 1977.

Hall, A. R. "Sir Isaac Newton's notebook, 1661–1665." *Cambridge Historical Journal* 9 (1948): 239.

—— *The Scientific Revolution 1500–1800: The Formation of the Modern Scientific Attitude.* London: Longmans, Green, 1954.

—— *From Galileo to Newton 1630–1720.* New York: Harper and Row, 1963.

Hall, A. R., and M. B. Hall. *Unpublished Scientific Papers of Isaac Newton.* New York: Cambridge University Press, 1962.

Halley, E. "Philosophiae naturalis principia mathematica." *Philosophical Transactions* 16 (1687): 291.

—— "Monsieur Cassini and his new and exact tables." *Philosophical Transactions* 18 (1694): 237.

—— "An account of several nebulae or lucid spots like clouds, lately discovered among the fixt stars by help of the telescope." *Philosophical Transactions* 29 (1716): 390.

—— "Considerations on the change of the latitudes of some of the principal fixt stars." *Philosophical Transactions* 30 (1717–1719): 736.

—— "Of the infinity of the sphere of fix'd stars." *Philosophical Transactions* 31 (1720–1721): 22.

—— "Of the number, order, and light of the fix'd stars." *Philosophical Transactions* 31 (1720–1721): 24.

—— *Correspondence and Papers of Edmund Halley.* Ed. E. F. MacPike. Oxford: Clarendon Press, 1932.

Haramundanis, K., ed. *Cecilia Payne-Gaposchkin: An Autobiography and Other Recollections.* New York: Cambridge University Press, 1984.

Harrison, E. R. "Visual acuity and the cone cell distribution of the retina." *British Journal of Ophthalmology* 37 (1953): 538.

—— "Olbers' paradox." *Nature* 204 (1964): 271.

—— "Olbers' paradox and the background radiation in an isotropic homogeneous universe." *Monthly Notices of the Royal Astronomical Society* 131 (1965): 1.

———— "Why the sky is dark at night." *Physics Today* 28 (1974): 69. Reprinted in *Astrophysics Today,* ed. A. G. W. Cameron. New York: American Institute of Physics, 1984.

———— "The dark night sky paradox." *American Journal of Physics* 45 (1977): 119.

———— "Radiation in homogeneous and isotropic models of the universe." *Vistas in Astronomy* 20 (1977): 341.

———— *Cosmology: The Science of the Universe.* New York: Cambridge University Press, 1981.

———— "The dark night-sky riddle: a 'paradox' that resisted solution." *Science* 226 (1984): 941. Reprinted in *Astronomy and Astrophysics,* ed. M. S. Roberts. Washington, D. C.: American Association for the Advancement of Science, 1985.

———— "Newton and the infinite universe." *Physics Today* 39 (1986): 24.

———— "Kelvin on an old and celebrated hypothesis." *Nature* 322 (1986): 417.

Haskins, C. H. *The Rise of Universities.* Ithaca, N.Y.: Cornell University Press, 1957.

Heath, T. L. *Aristarchus of Samos: The Ancient Copernicus.* New York: Dover Publications, 1981.

Heninger, S. K. *The Cosmographical Glass: Renaissance Diagrams of the Universe.* San Marino, California: Huntington Library, 1977.

Herschel, J. F. W. *A Treatise of Astronomy.* London: Longmans, 1830.

———— *A Preliminary Discourse on the Study of Natural Philosophy.* Philadelphia: Carey and Lea, 1831.

———— "Humboldt's *Kosmos.*" *The Edinburgh Review* 87 (1848): 170. Reprinted in *Essays,* 1857, p. 257.

———— *Outlines of Astronomy.* London: Longmans, Green, 1849.

———— "Treatises on sound and light." In *Encyclopaedia Metropolitana.* London: Richard Griffin, 1854.

———— *Essays from the Edinburgh and Quarterly Reviews with Addresses and Other Pieces.* London: Longman, Brown, Green, Longmans & Roberts, 1857.

———— *Herschel at the Cape: Diaries and Correspondence of Sir John Herschel, 1834–1838.* Ed. D. S. Evans, T. J. Deeming, B. H. Evans, and S. Goldfarb. Austin: University of Texas Press, 1969.

Herschel, W. "On the construction of the heavens." *Philosophical Transactions* 75 (1785): 213.

———— "Catalogue of 500 new nebulae, nebulous stars, planetary nebulae, and clusters of stars; with remarks on the construction of the heavens." *Philosophical Transactions* 92 (1802): 477.

———— *The Scientific Papers of Sir William Herschel.* Ed. J. L. E. Dreyer. 2 vols. London: Royal Society and Royal Astronomical Society, 1912.

Hesse, M. B. *Force and Fields: The Concept of Action at a Distance in the History of Physics.* London: Nelson and Sons, 1961.

Hoagland, C. "The universe of *Eureka:* a comparison of the theories of Eddington and Poe." *Southern Literary Messenger* 1 (1939): 307.

Hoffman, B. *Albert Einstein: Creator and Rebel*. New York: Viking, 1972.

Holden, E. S. *Sir William Herschel: His Life and Works*. New York: Scribner's Sons, 1881.

Holton, G. "Johannes Kepler's universe: its physics and metaphysics." *American Journal of Physics* 24 (1956): 340.

Hooke, R. *Micrographia: Or Some Physiological Descriptions of Minute Bodies Made by Magnifying Glasses with Observations and Inquiries Thereupon*. London: Royal Society, 1665. Reprinted, New York: Dover Publications, 1961.

—— *The Posthumous Works of Robert Hooke, Containing His Cutlerian Lectures, and Other Discourses*. Ed. R. Waller. 1705. Rev. ed., R. S. Westfall. New York: Johnson Reprint Corporation, 1969.

—— *Philosophical Experiments and Observations of the Late Eminent Dr. Robert Hooke*. Ed. W. Derham. London: Frank Cass, 1967.

Hoskin, M. A. *William Herschel and the construction of the Heavens*. New York: Norton, 1963.

—— "The cosmology of Thomas Wright of Durham." *Journal for the History of Astronomy* 1 (1970): 44.

—— "Dark skies and fixed stars." *Journal of the British Astronomical Association* 83 (1973): 254.

—— "The 'great debate': what really happened." *Journal for the History of Astronomy* 7 (1976): 169.

—— "Newton, providence and the universe of stars." *Journal for the History of Astronomy* 8 (1977): 77.

—— "The English background to the cosmology of Wright and Herschel." In *Cosmology, History, and Theology*, ed. W. Yougrau and A. D. Breck. New York: Plenum Press, 1977.

—— "Stukeley's cosmology and the Newtonian origins of Olbers's paradox." *Journal for the History of Astronomy* 16 (1985): 77.

Hoyle, F. "A new model for the expanding universe." *Monthly Notices of the Royal Astronomical Society* 108 (1948): 372.

—— *Frontiers in Astronomy*. London: Heinemann, 1955.

Hubble, E. "Cepheids in spiral nebulae." *Publications of the American Astronomical Society* 5 (1925): 261.

—— "Extra-galactic nebulae." *Astrophysical Journal* 64 (1926): 321.

—— "A relation between distance and radial velocity among extra-galactic nebulae." *Proceedings of the National Academy of Sciences* 15 (1929): 168.

—— *The Realm of the Nebulae*. New Haven: Yale University Press, 1936.

—— *The Observational Approach to Cosmology*. Oxford: Oxford University Press, 1937.

Hufbauer, K. "Astronomers take up the stellar-energy problem, 1917–1920." *Historical Studies in the Physical Sciences* 11 (1981): 277.

Huggins, W. "On the spectrum of the Great Nebula in Orion." *Monthly Notices of the Royal Astronomical Society* 25 (1865): 155.

—— "Further observations on the spectra of some of the stars and nebulae, with an attempt to determine therefrom whether these bodies are moving towards or from the Earth, also observations on the spectra of the Sun and comets II." *Philosophical Transactions* 158 (1868): 529.

——— "On the spectrum of the Great Nebula in Orion, and on the motions of some stars towards or from the Earth." *Proceedings of the Royal Society* 20 (1872): 379.

——— "On the photographic spectra of stars." *Proceedings of the Royal Society* 25 (1876): 445.

——— "Celestial spectroscopy." *British Association* 61 (1891): 3. Reprinted in *Smithsonian Report* 43 (1891): 69.

——— "The new astronomy: a personal retrospect." *The Nineteenth Century* 41 (1897): 907.

Huggins, Sir William, and Lady Margaret Huggins. *Publications of Sir William Huggins's Observatory.* Vol. 1, *An Atlas of Representtive Stellar Spectra.* London: Wesley, 1899. Vol. 2, *The Scientific Papers of Sir William Huggins.* London: Wesley, 1909.

Huggins, W., and W. A. Miller. "Note on the lines in the spectra of some of the fixed stars." *Proceedings of the Royal Society* 12 (1863): 444.

——— "On the spectra of some of the fixed stars." *Philosophical Transactions* 154 (1864): 412.

Hujer, K. "Sesquicentennial of Christian Doppler." *American Journal of Physics* 23 (1955): 51.

Humboldt, A. von. *Kosmos.* 5 vols. Stuttgart: J. G. Cotta, 1845–1862. Trans. E. C. Otte. 5 vols. London: H. G. Bohn, 1848–1865.

Huygens, C. *Traité de la Lumière* (Treatise on Light). 1690. Trans. S. P. Thompson. London: Macmillan, 1912. Reprinted in *Great Books of the Western World,* vol. 34. Chicago: Encyclopaedia Britannica, 1955, p. 551.

——— *Cosmotheros* and *The Celestial Worlds Discover'd* 1698. Reprinted, London: Frank Cass, 1968.

Hyland, D. A. *The Origins of Philosophy: Its Rise in Myth and the Pre-Socratics.* New York: Putnam, 1973.

Jacob, M. C. *The Newtonians and the English Revolution, 1689–1720.* Ithaca, N.Y.: Cornell University Press, 1976.

Jaki, S. L. "Olbers', Halley's, or whose paradox?" *American Journal of Physics* 35 (1967): 200.

——— *The Paradox of Olbers' Paradox.* New York: Herder and Herder, 1969.

——— "New light on Olbers' dependence on Chéseaux." *Journal for the History of Astronomy* 1 (1970): 53.

——— *The Milky Way: An Elusive Road for Science.* New York: Science History Publications, 1972.

——— *Planets and Planetarians: A History of Theories of the Origin of Planetary Systems.* New York: Wiley, 1979.

Jammer, M. *Concepts of Force.* Cambridge, Mass.: Harvard University Press, 1957.

——— *Concepts of Space: The History of the Theories of Space in Physics.* New York: Harper, 1960.

Jaspers, K. *Anselm and Nicholas of Cusa.* New York: Harcourt Brace Jovanovich, 1966.

Jeans, J. H. "The stability of a spherical nebula." *Philosophical Transactions* 199 (1902): 48.

———— *Astronomy and Cosmogony.* Cambridge: Cambridge University Press, 1929.

Johnson, F. R. *Astronomical Thought in Renaissance England: A Study of the English Scientific Writings from 1500 to 1645.* Baltimore: Johns Hopkins Press, 1937.

———— "Gresham College: precursor of the Royal Society." *Journal of the History of Ideas* 1 (1940): 413.

———— "Thomas Digges and the infinity of the universe." In *Theories of the Universe,* ed. M. K. Munitz. New York: The Free Press, 1957.

Kahn, C. H. *Anaximander and the Origin of Greek Cosmology.* New York: Columbia University Press, 1960.

Kant, I. *Kant's Cosmogony.* Trans. W. Hastie. Glasgow: Maclehose, 1900.

———— *Kant's Cosmogony.* Trans. W. Hastie. Introduction W. Ley. New York: Greenwood, 1968.

———— *Universal Natural History and Theory of the Heavens.* Trans. W. Hastie. Introduction M. K. Munitz. Ann Arbor: University of Michigan Press, 1969.

———— *Kant's Cosmogony: As in His Essay on the Retardation of the Rotation of the Earth and His Natural History and Theory of the Heavens.* Trans. W. Hastie. Introduction G. J. Whitrow. New York: Johnson Reprint Corporation, 1970.

Kargon, R. *Atomism in England from Hariot to Newton.* Oxford: Clarendon Press, 1966.

Kayser, H. "Scientific worthies—Sir William Huggins." *Nature* 69 (1901): 225.

Kellogg, O. D. *Foundations of Potential Theory.* New York: Dover Publications, 1953.

Kelvin. See W. Thompson.

Kepler, J. *Johannes Kepler Gesammelte Werke.* Ed. W. von Dyck and M. Caspar. 15 vols. Munich: C. H. Beck'sche Verlagsbuchhandlung, 1937. Vol. 1, *Mysterium Cosmographicum* (1596), *De Stella Nova* (1606); Vol. 6, *Harmonices Mundi* (1619); Vol. 7, *Epitome Astronomiae Copernicanae* (1618).

———— *Kepler's Conversation with Galileo's Sidereal Messenger.* Trans. E. Rosen. New York: Johnson Reprint Corporation, 1965.

———— *Mysterium Cosmographicum: The Secret of the Universe.* Trans. A. M. Duncan. Introduction E. J. Aiton. New York: Abaris Books, 1981.

———— "The discovery of the laws of planetary motion." Trans. J. H. Walden from *Harmonices Mundi.* In *A Source book in Astronomy,* ed. H. Shapley and H. E. Howarth. New York: McGraw-Hill, 1929.

King, H. C. *The History of the Telescope.* London: Charles Griffin, 1955.

Kirchhoff, G. "On Fraunhofer's lines." Trans G. G. Stokes. *Philosophical Magazine* 19 (1860): 195.

———— "Contributions towards the history of spectrum analysis and of the analysis of the solar atmosphere." *Philosophical Magazine* 25 (1863): 250.

Kirchhoff, G., and R. Bunsen. "Chemical analysis by spectrum observations." *Philosophical Magazine* 20 (1960): 89.

Kirk, G. S., and J. E. Raven. *The Presocratic Philosophers.* New York: Cambridge University Press, 1966.

Klein, M. J. "Maxwell, his demon, and the second law of thermodynamics." *American Scientist* 58 (1970): 84.

M. Kline. *Mathematical Thought from Ancient to Modern Times.* New York: Oxford University Press, 1972.

Koestler, A. *The Watershed: A Biography of Johannes Kepler.* Garden City, N.Y.: Doubleday, 1960.

Koyré, A. *From the Closed World to the Infinite Universe.* Baltimore: Johns Hopkins Press, 1957.

——— *Newtonian Studies.* Cambridge, Mass.: Harvard University Press, 1965.

——— *The Astronomical Revolution: Copernicus, Kepler, Borelli.* Trans. R. E. W. Maddison. Ithaca, N.Y.: Cornell University Press, 1973.

——— *Galileo Studies.* Trans. J. Mepham. Atlantic Highlands, N.J.: Humanities Press, 1978.

Kubrin, D. "Newton and the cyclical cosmos: providence and the mechanical philosophy." *Journal of the History of Ideas* 28 (1967): 325.

Kuhn, T. S. *Copernican Revolution: Planetary Astronomy in the Development of Western Thought.* Cambridge, Mass.: Harvard University Press, 1957.

Lambert, J. H. *Cosmological Letters on the Arrangement of the World-Edifice.* Trans. S. L. Jaki. New York: Science History Publications, 1976.

Lanczos, C. *Albert Einstein and the Cosmic World Order.* New York: Interscience, 1965.

Langley, S. P. *The New Astronomy.* Boston: Ticknor, 1888.

Laplace, P. S. de. *Celestial Mechanics.* Trans. N. Bowditch. 4 vols. New York: Chelsea, 1966.

Lazer, D. "The significance of Newtonian cosmology." *Astronomical Journal* 59 (1954): 168.

Lemaître, G. "A homogeneous universe of constant mass and increasing radius accounting for the radial velocity of extra-galactic nebulae." *Monthly Notices of the Royal Astronomical Society* 91 (1931): 483.

——— *The Primeval Atom: An Essay on Cosmogony.* Trans. B. H. Korff and S. A. Korff. New York: Van Nostrand, 1951.

Lewis, C. S. *The Discarded Image: An Introduction to Medieval and Renaissance Literature.* Cambridge: Cambridge University Press, 1967.

Lindberg, D. C., ed. *A Source Book in Medieval Science.* Cambridge, Mass.: Harvard University Press, 1974.

——— *Theories of Vision from al-Kindi to Kepler.* Chicago: University of Chicago Press, 1976.

——— "The science of optics." In *Science in the Middle Ages,* ed. D. C. Lindberg. Chicago: University of Chicago Press, 1978.

————, ed. *Science in the Middle Ages.* Chicago: University of Chicago Press, 1978.

Locke, J. *An Essay Concerning Human Understanding.* Ed. P. H. Nidditch. Oxford: Clarendon Press, 1975.

Lodge, O. *The Ether of Space.* New York: Harper, 1909.

———— *Ether and Reality.* New York: Doran, 1925.

Lovejoy, A. O. *The Great Chain of Being: A Study of the History of an Idea.* Cambridge, Mass.: Harvard University Press, 1936.

Lubbock, C. A. *The Herschel Chronicle.* Cambridge: Cambridge University Press, 1933.

Lucretius. *Lucretius on the Nature of Things.* Trans. C. Bailey. Oxford: Clarendon Press, 1922.

———— *Lucretius, with an English Translation.* Trans. W. H. D. Rouse. Cambridge, Mass.: Loeb Classical Library, 1924.

———— *T. Lucreti Cari: De Rerum Natura.* Ed. W. E. Leonard and S. B. Smith. Madison: University of Wisconsin Press, 1942.

———— *Titi Lucreti Cari: De Rerum Natura.* Trans. C. Bailey. Oxford: Clarendon Press, 1947.

———— *The Nature of the Universe.* Trans. R. E. Latham. Harmondsworth: Penguin, 1951.

MacDonald, D. K. C. *Faraday, Maxwell, and Kelvin.* Garden City, N.Y.: Doubleday, 1964.

MacMillan, W. D. "On stellar evolution." *Astrophysical Journal* 48 (1918): 35.

———— "Some postulates of cosmology." *Scientia* 31 (1922): 105.

———— "Some mathematical aspects of cosmology." *Science* 62 (1925): 63, 96, 121.

MacPike, E. F. *Correspondence and Papers of Edmond Halley.* London: Taylor and Francis, 1937.

Maddison, R. E. W. *The Life of the Honorable Robert Boyle.* London: Taylor and Francis, 1969.

Mandelbrot, B. B. *The Fractal Geometry of Nature.* San Francisco: W. H. Freeman, 1982.

Manning, H. P. *The Fourth Dimension Simply Explained.* New York: Dover Publications, 1960.

Manuel, F. E. *Isaac Newton, Historian.* Cambridge, Mass.: Harvard University Press, 1963.

———— *A Portrait of Isaac Newton.* Cambridge, Mass.: Harvard University Press, 1968.

———— *The Religion of Isaac Newton.* Oxford: Clarendon Press, 1974.

Marsak, L. M. "Cartesianism in Fontenelle and French Science, 1686–1752." *Isis* 50 (1959): 51.

Maxwell, J. C. "On physical lines of force. Part II. The theory of molecular vortices applied to electric currents." *Philosophical Magazine* 21 (1861): 281.

———— *The Scientific Papers of James Clerk Maxwell.* Ed. W. D. Niven. 2 vols. Cambridge: Cambridge University Press, 1890.

McColley, G. "The seventeenth century doctrine of a plurality of worlds." *Annals of Science* 1 (1936): 385.

McCrea, W. H. "On the significance of Newtonian cosmology." *Astronomical Journal* 60 (1955): 2718.

———— "James Bradley 1693–1762." *Quarterly Journal of the Royal Astronomical Society* 4 (1962): 38.

———— "Willem de Sitter, 1872–1934." *Quarterly Journal of the Royal Astronomical Association* 82 (1972): 178.

McCrea, W. H., and E. Milne. "Newtonian universe and the curvature of space." *Quarterly Journal of Mathematics* 5 (1934): 73.

McGucken, W. *Nineteenth-Century Spectroscopy: Development of Understanding of Spectra 1802–1897.* Baltimore: Johns Hopkins Press, 1969.

McGuire, J. E., and M. Tamny. *Certain Philosophical Questions: Newton's Trinity Notebook.* New York: Cambridge University Press, 1983.

McVittie, G. C., and S. P. Wyatt. "Background radiation in a Milne universe." *Astrophysical Journal* 130 (1959): 1.

Meadows, A. J. *Early Solar Physics.* Oxford: Oxford University Press, 1970.

———— "The origins of astrophysics." *American Scientist,* May–June 1984, p. 269.

Meyer, K. "Ole Roemer and the thermometer." *Nature* 137 (1910): 296.

Michel, P. H. *The Cosmology of Giordano Bruno.* Trans. R. E. W. Maddison. Ithaca, N.Y.: Cornell University Press, 1973.

Middleton, W. E. K. *The History of the Barometer.* Baltimore: Johns Hopkins Press, 1964.

Miller, D. G. "Ignored intellect: Pierre Duhem." *Physics Today* 19 (1966): 47.

Miller, P. "Newton's four letters to Bentley, and the Boyle Lectures related to them." In *Isaac Newton's Papers and Letters on Natural Philosophy, and Related Documents,* ed. I. B. Cohen and R. G. Schofield. Cambridge, Mass.: Harvard University Press, 1978.

Mills, C. E., and C. F. Brooke, eds. *A Sketch of the Life of Sir William Huggins.* London: 1936.

Milne, E. "A Newtonian expanding universe." *Quarterly Journal of Mathematics* 5 (1934): 64.

Minnaert, M. *The Nature of Light and Colour in the Open Air.* Trans. H. M. Kremer-Priest. Rev. K. E. B. Jay. New York: Dover Publications, 1954.

Munitz, M. K., ed. *Theories of the Universe: From Babylonian Myth to Modern Science.* New York: Free Press, 1957.

Neugebauer, O. *The Exact Sciences in Antiquity.* New York: Dover Publications, 1969.

Neumann, C. *Untersuchungen über das Newton'sche Prinzip der Fernwirkung.* Leipzig: Teubner, 1896.

Newcomb, S. "Elementary theorems relating to the geometry of a space of three dimensions and of uniform positive curvature in the fourth dimension." *Journal für die Reine and Ungewandte Mathematik* 83 (1877): 293.

—— *Popular Astronomy.* New York: Harper, 1878.

—— *The Stars: A Study of the Universe.* London: Putnam's Sons, 1901.

—— *Astronomy for Everybody.* New York: McClure, Phillips, 1902.

Newton, I. "The Optical lectures, 1670–1672." In *The Optical Papers of Isaac Newton,* vol. 1, ed. A. E. Shapiro. New York: Cambridge University Press, 1983.

—— *Isaac Newton's Mathematical Principles of Natural Philosophy and His System of the World.* 2nd ed. Trans. A. Motte. 1729. Rev. F. Cajori. Berkeley: University of California Press, 1934.

—— *The Mathematical Principles of Natural Philosophy.* Trans. A. Motte. Introduction I. B. Cohen. London: Dawsons, 1968.

—— *Opticks or a Treatise of the Reflections, Refractions, Inflections & Colours of Light.* 4th ed. 1730. Foreword A. Einstein. Introduction E. Whittaker. Preface I. B. Cohen. New York: Dover Publications, 1952.

—— *The Chronology of Ancient Kingdoms Amended.* London: 1728.

—— *Unpublished Scientific Papers of Isaac Newton.* Ed. A. R. Hall and M. B. Hall. Cambridge: Cambridge University Press, 1962.

—— *A Treatise of the System of the World.* Trans. I. B. Cohen. London: Dawsons, 1969.

—— *The Correspondence of Isaac Newton.* Ed. H. W. Turnbull, J. F. Scott, A. R. Hall, and L. Tilling. 7 vols. Cambridge: Cambridge University Press, 1959–1977.

—— *The Mathematical Papers of Isaac Newton.* Ed. D. T. Whiteside. 8 vols. Cambridge: Cambridge University Press, 1967–1981.

—— *Isaac Newton's Papers and Letters on Natural Philosophy and Related Documents.* Ed. I. B. Cohen and R. S. Schofield. Cambridge, Mass.: Harvard University Press, 1958.

Nichol. J. P. *Views of the Architecture of the Heavens. In a Series of Letters to a Lady.* Edinburgh, 1838. New York: Dayton and Newman, 1842.

Nicolson, M. N. "The early stage of Cartesianism in England." *Studies in Philology* 26 (1929): 356.

—— "The new astronomy and English imagination." *Studies in Philology* 32 (1935): 428.

—— "The telescope and imagination." *Modern Philology* 32 (1935): 233.

—— *The Breaking of the Circle: Studies in the Effect of the "New Science" upon Seventeenth Century Poetry.* Evanston, Ill.: Northwestern University Press, 1950.

—— *Science and the Imagination.* Ithaca, N.Y.: Cornell University Press, 1956.

—— *Mountain Gloom and Mountain Glory.* Ithaca, N.Y.: Cornell University Press, 1959.

—— *Pepys' Diary and the New Science.* Charlottesville: University Press of Virginia, 1965.

North, J. D. *The Measure of the Universe: A History of Modern Cosmology.* Oxford: Clarendon Press, 1965.

———— "Chronology and the age of the world." In *Cosmology, History, and Theology,* ed. W. Yougrau and A. D. Breck. New York: Plenum Press, 1977.

Numbers, R. L. *Creation by Natural Law: Laplace's Nebula Hypothesis in American Thought.* Seattle: University of Washington Press, 1977.

Oates, W. J. *The Stoic and Epicurean Philosophers: The Complete Extant Writings of Epicurus, Epictetus, Lucretius, Marcus Aurelius.* New York: Random House, 1940.

Olbers, H. W. M. "Ueber die Durchsichtigkeit des Weltraumes." In *Astronomisches Jahrbuch für das Jahr 1826,* ed. J. E. Bode. Berlin: Späthen;, 1823. Trans. "On the transparency of space." *Edinburgh New Philosophical Journal* 1 (1826): 141.

Orchard, T. N. *Milton's Astronomy: The Astronomy of 'Paradise Lost.'* New York: Longmans, Green, 1913.

Orr, M. A. *Dante and the Early Astronomers.* London: Allan Wingate, 1956.

Pancheri, L. V. "Pierre Gassendi, a forgotten but important man in the history of physics." *American Journal of Physics* 46 (1978): 455.

Paneth, F. A. "Thomas Wright of Durham." *Endevour* 9 (1950): 117.

Parsons, C., ed. *The Scientific Papers of William Parsons, Third Earl of Rosse.* London: Percy Lund, 1926.

Partington, J. R. "The origins of the atomic theory." *Annals of Science* 4 (1939): 245.

Paterson, A. M. *The Infinite Worlds of Giordano Bruno.* Springfield, Ill.: Charles C. Thomas, 1970.

Patrizi, F. "On physical space." Trans. B. Brickman. *Journal of the History of Ideas* 4 (1943): 224.

Pegg, D. T. "Night sky darkness in the Eddington-Lemaître universe." *Monthly Notices of the Royal Astronomical Society* 154 (1971): 321.

Penzias, A. A., and R. W. Wilson. "A measurement of excess antenna temperature at 4080 MHz." *Astrophysical Journal* 142 (1965): 419.

Pepys, S. *The Diary of Samuel Pepys.* Ed. R. Latham and W. Mathews. Berkeley: University of California Press, 1872.

Peterson, M. A. "Dante and the 3-sphere." *American Journal of Physics* 47 (1980): 1031.

Plotkin, H. "Henry Draper, Edward C. Pickering, and the birth of American astrophysics." *Annals of the New York Academy of Sciences* 395 (1982): 321.

Plutarch. *Plutarch's Essays and Miscellania.* Ed. W. W. Goodwin. Boston: Little Brown, 1906.

Poe, E. A. "The power of words." *United States Magazine and Democratic Review,* June 1845.

———— *Eureka: A Prose Poem.* New York: Putnam, 1848.

———— *Eureka: A Prose Poem.* Ed. R. P. Benton. Hartford, Conn.: Transcendental Books, 1973.

—— *The Science Fiction of Edgar Allan Poe.* Ed. H. Beaver. Harmondsworth: Penguin, 1976.

Priestley, J. *The History and Present State of Discoveries Relating to Vision, Light, and Colours.* London: Johnson, 1772.

Proctor, R. A. *Other Worlds than Ours: The Plurality of Worlds Studied under the Light of Recent Scientific Researches.* London: Longmans, Green, 1870.

—— *The Expanse of Heaven: A Series of Essays on the Wonders of the Firmament.* New York: Appleton, 1874.

—— *Our Place among the Infinities: A Series of Essays Contrasting Our Little Abode in Space and Time with the Infinities around Us.* London: King, 1876.

—— *Other Suns than Ours: A Series of Essays on Suns—Old, Young, and Dead.* London: Longmans, Green, 1896.

Quinn, A. H. *Edgar Allan Poe: A Critical Biography.* New York: Appleton Century, 1941.

Richardson, R. S. "Astronomy—the distaff side." *Astronomical Society of the Pacific.* Leaflet no. 181, March 1944.

—— "Lady Huggins and others." In *The Star Lovers.* New York: Macmillan, 1967.

—— "Edmund Halley: to fix the 'frame of the world.'" In *The Star Lovers.* New York: Macmillan, 1967.

Rindler, R. "Visual horizons in world-models." *Monthly Notices of the Royal Astronomical Society* 116 (1956): 662.

Rist, J. M. *Stoic Philosophy.* Cambridge: Cambridge University Press, 1969.

Roemer, O. "A demonstration concerning the motion of light, communicated from Paris, in the *Journal des Scavans,* and here made English." *Philosophical Transactions* 11 (1677): 893.

Ronan, C. A. *Edmund Halley: Genius in Eclipse.* Garden City, N.Y.: Doubleday, 1969.

Ronchi, V. *The Nature of Light: A Historical Survey.* Trans. V. Barocas. Cambridge, Mass.: Harvard University Press, 1970.

Rosen, E. *The Naming of the Telescope.* New York: Henry Schuman, 1947.

—— "The title of Galileo's 'Sidereus nuncius.'" *Isis* 41 (1950): 287.

—— "Galileo and the telescope." *Scientific Monthly* 72 (1951): 180.

—— "The invention of eyeglasses." *Journal of the History of Medicine* 11 (1956): 13.

—— *Kepler's Conversation with Galileo's Sidereal Messenger.* New York: Johnson Reprint Corporation, 1965.

Rowan-Robertson, M. *The Cosmological Distance Ladder.* New York: W. H. Freeman, 1985.

Russell, A. *Lord Kelvin: His Life and Work.* London: Jack, 1912.

Russell, B. *Mysticism and Logic.* London: Allen and Unwin, 1917.

Russell, H. N., R. S. Dugan, and J. Q. Stewart. *Astronomy.* 2 vols. New York: Ginn, 1955.

Russell, J. L. "English astronomy before 1675." *Nature* 255 (1975): 583.

Sabra, A. I. *Theories of Light from Descartes to Newton.* London: Oldbourne, 1967. Reprinted, New York: Cambridge University Press, 1981.

Salmon, W. C. *Zeno's Paradoxes.* Indianapolis: Bobbs-Merrill, 1970.

Sambursky, S. *Physics of the Stoics.* London: Routledge and Kegan Paul, 1959.

—— *The Physical World of Late Antiquity.* New York: Basic Books, 1962.

—— *The Physical World of the Greeks.* London: Routledge and Kegan Paul, 1963.

Sandbach, F. H. *The Stoics.* London: Chatto and Windus, 1975.

Sanders, J. H. *Velocity of Light.* London: Pergamon, 1965.

Sarton, G. *Introduction to the History of Science.* Baltimore: Johns Hopkins Press, 1927.

—— "Discovery of the aberration of light." *Isis* 16 (1931): 233.

Saunders, J. L. *Greek and Roman Philosophy after Aristotle.* New York: Free Press, 1966.

Schaffer, S. "The phoenix of nature: fire and evolutionary cosmology in Wright and Kant." *Journal for the History of Astronomy* 9 (1978): 180.

Schaffner, K. F. *Nineteenth-Century Aether Theories.* New York: Pergamon, 1972.

Schlegel, R. "Steady-state theory at Chicago." *American Journal of Physics* 26 (1958): 601.

Schwarzschild, K. "Ueber das zulässige Krümmungsmass des Raumes." *Vierteljahrschrift der Astronomischen Gesellschaft* 35 (1900): 337.

Schwarzschild, M. *Structure and Evolution of the Stars.* Princeton: Princeton University Press, 1958.

Sciama, D. W. *The Unity of the Universe.* London: Faber and Faber, 1959.

—— *Modern Cosmology.* Cambridge: Cambridge University Press, 1971.

Scott, J. F. *The Scientific Work of René Descartes.* London: Taylor and Francis, 1952.

Seeliger, H. "Ueber das Newton'sche Gravitationsgesetz." *Astronomische Nachrichten* 137, no. 3273 (1895).

—— "Ueber das Newton'sche Gravitationsgesetz." *Sitzungsberichte der Mathematisch-Physicalischen Classe, Akademie der Wissenschaften, München* 26 (1896): 373.

Shapiro, H. *Motion, Time, and Place According to William Ockham.* New York: Franciscan Institute, 1957.

Shapley, H. "Colors and magnitudes in stellar clusters. Second part: Thirteen hundred stars in the Hercules Cluster (Messier 13)." *Astrophysical Journal* 45 (1917): 123.

—— "Evolution of the idea of galactic size." *Bulletin of the National Research Council of the National Academy of Sciences* 2 (1921): 171.

—— *Flights from Chaos.* New York: McGraw-Hill, 1930.

—— ed. *Source Book in Astronomy 1900–1950.* Cambridge, Mass.: Harvard University Press, 1960.

—— *Through Rugged Ways to the Stars.* New York: Scribner's Sons, 1969.

Shapley, H. and H. E. Howarth, eds. *A Source Book in Astronomy.* New York: McGraw-Hill, 1929.

Sharlin, H. I., and T. Sharlin. *Lord Kelvin: The Dynamic Victorian.* University Park: Pennsylvania State University Press, 1979.

Sheldon-Williams, I. P. "The pseudo-Dionysius." In *The Cambridge History of Later Greek and Early Medieval Philosophy.* Ed. A. H. Armstrong. Cambridge: Cambridge University Press, 1967.

Sidgwick, J. B. *William Herschel: Explorer of the Heavens.* London: Faber, 1953.

Singer, C. "The scientific views and visions of Saint Hildegard (1098–1180)." In *Studies in the History and Method of Science,* ed. C. Singer. London: Dawson, 1955.

Singer, D. W. *Giordano Bruno: His Life and Thought, with an Annotated Translation of His Work on the Infinite Universe and Worlds.* New York: Schumann, 1950.

Smith, R. W. *The Expanding Universe: Astronomy's 'Great Debate' 1900–1931.* Cambridge: Cambridge University Press, 1982.

Solmson, F. *Aristotle's System of the Physical World.* Ithaca, N.Y.: Cornell University Press, 1960.

Spurgeon, C. F. E. *Shakespeare's Imagery.* New York: Macmillan, 1935.

Struve, O. "The constitution of diffuse matter in interstellar space." *Journal of the Washington Academy of Science* 31 (1941): 217.

———— "Some thoughts on Olbers' paradox." *Sky and Telescope* 25 (1963): 140.

Struve, O., and V. Zebergs. *Astronomy in the 20th Century.* New York: Macmillan, 1962.

Stukeley, W. *Memoirs of Sir Isaac Newton's Life, 1752: Being Some Account of His Family and Chiefly of the Junior Part of His Life.* Ed. A. H. White. London: Taylor and Francis, 1936.

Swedenborg, E. *Principia Rerum Naturalium.* Trans. J. R. Rendell and I. Tansley. London: Swedenborg Society, 1912.

Tammann, G. A. "Jean-Philippe de Loys de Chéseaux and his discovery of the so-called Olbers' paradox." *Scientia* 60 (1966): 22.

Taylor, F. S. *An Illustrated History of Science.* Illustrated A. R. Thomson. London: Heinemann, 1955.

Terzian, Y., and E. M. Bilson, eds. *Cosmology and Astrophysics: Essays in Honor of Thomas Gold.* Ithaca, N.Y.: Cornell University Press, 1982.

Thompson, S. P. *Michael Faraday, His Life and Work.* London: Cassell, 1901.

———— "Lord Kelvin." *Nature* 77 (1907): 175.

———— *The Life of William Thomson, Baron Kelvin of Largs.* 2 vols. London: Macmillan, 1910.

Thomson, J. *The Complete Poetical Works of James Thomson.* Ed. J. Robertson. London: Oxford University Press, 1908.

Thomson, W. (Lord Kelvin). "On a mechanical representation of electric, magnetic, and galvanic forces." *Cambridge and Dublin Mathematical Journal* 2 (1847): 61.

———— "Note on the possible density of the luminiferous medium, and on the mechanical value of a cubic mile of sunlight." *Edinburgh Royal Society Transactions* 21, pt. 1 (May 1854).

———— *Notes of Lectures on Molecular Dynamics and the Wave Theory of Light,* reported by A. S. Hathaway. Baltimore: Johns Hopkins University Press, 1884.

———— *Popular Lectures and Addresses.* 3 vols. London: Macmillan, 1891–1894.

———— "On ether and gravitational matter through infinite space." *Philosophical Magazine* 2 (1901): 161.

———— "On the clustering of gravitational matter in any part of the universe." *Nature* 64 (1901): 626.

———— "On the clustering of gravitational matter in any part of the universe." *Philosophical Magazine* 3 (1902): 1.

———— "Lord Kelvin and his first teacher in natural philosophy." *Nature* 68 (1903): 623.

———— *Mathematical and Physical Papers.* Ed. J. Larmor. 6 vols. Cambridge: Cambridge University Press, 1882–1911.

———— *Baltimore Lectures on Molecular Dynamics and the Wave Theory of Light.* Cambridge: Cambridge University Press, 1904.

———— "William Thomson, Baron Kelvin of Largs (1824–1907)" (obituary by J. Larmor). *Proceedings of the Royal Society* 81 (1908): iii.

Thorndike, L. *A History of Magic and Experimental Science.* New York: Macmillan, 1923.

Tillyard, E. M. W. *The Elizabethan World Picture.* New York: Macmillan, 1944.

Tolstoy, I. *James Clerk Maxwell, a Biography.* Edinburgh: Canongate, 1981.

Toulmin, S., and J. Goodfield. *The Fabric of the Heavens: The Development of Astronomy and Dynamics.* New York: Harper and Brothers, 1961.

Trumpler, R. J. "Preliminary results on the distances, dimensions, and space distribution of open clusters." *Lick Observatory Bulletin* 14, no. 420 (1930): 154–188.

Turnbull, H. W., ed. *James Gregory: Tercentenary Memorial Volume.* London: Bell, 1939.

Twain, M. (S. L. Clemens). *The Science Fiction of Mark Twain,* ed. D. Ketterer. New York: Archon Books, 1984.

Van Helden, A. "The telescope in the seventeenth century. *Isis* 65 (1974): 38.

———— "The invention of the telescope." *Transaction of the American Philosophical Society* 67, pt. 4 (1977).

———— "Roemer's speed of light." *Journal for the History of Astronomy* 15 (1983): 137.

———— *Measuring the Universe: Cosmic Dimensions from Aristarchus to Halley.* Chicago: University of Chicago Press, 1985.

Vehrenberg, H. *Atlas of Deep-Sky Splendors.* Cambridge, Mass.: Sky Publishing Corporation, 1978.

Voltaire. *Letters Concerning the English Nation.* London: Davis and Lyon, 1733.

———— *The Elements of Sir Isaac Newton's Philosophy.* Trans. J. Hanna. London: Stephen Austin, 1738. Reprinted, London: Frank Cass, 1967.

Vrooman, J. R. *René Descartes: A Biography.* New York: Putnam's Sons, 1970.

Waerden, B. L. van der. *Science Awakening.* Groningen, Holland: Noordhoff, 1954.

Wallace, A. R. *Man's Place in the Universe: A Study of the Results of Scientific Research in Relation to the Unity or Plurality of Worlds.* New York: McClure, Phillips, 1903.

Waterfield, R. L. *A Hundred Years of Astronomy.* London: Duckworth, 1938.

Weale, R. A. *From Sight to Light.* London: Oliver and Boyd, 1968.

Webster, C. "Henry More and Descartes: some new sources." *British Journal for the History of Science* 4 (1969): 359.

Weinberg, J. R. *A Short History of Medieval Philosophy.* Princeton: Princeton University Press, 1964.

Weinberg, S. *The First Three Minutes: A Modern View of the Origin of the Universe.* New York: Basic Books, 1977.

Westfall, R. S. *Science and Religion in Seventeenth-Century England.* New Haven: Yale University Press, 1958.

———— "The foundations of Newton's philosophy of nature." *British Journal for the History of Science* 1 (1962): 171.

———— *Force in Newton's Physics: The Science of Dynamics in the Seventeenth Century.* New York: Elsevier, 1971.

———— *Never at Rest: A Biography of Isaac Newton.* New York: Cambridge University Press, 1980.

Wheelwright, P. *Aristotle: Containing Selections from Seven of the Most Important Books of Aristotle.* New York: Odyssey Press, 1935.

Whewell, W. *History of the Inductive Sciences from the Earliest to the Present Times.* 3 vols. London: Parker, 1837.

White, L. *Medieval Technology and Social Change.* New York: Oxford University Press, 1962.

———— "Cultural climates and technological advance in the Middle Ages." *Viator* 2 (1871): 171.

Whiteside, D. T. "The expanding world of Newtonian research." *History of Science* 1 (1962): 16.

Whitney, C. A. *The Discovery of Our Galaxy.* New York: Knopf, 1971.

Whitrow, G. J. *The Structure and Evolution of the Universe.* London: Hutchinson, 1959.

———— "Why is the sky dark at night?" *History of Science* 10 (1971): 128.

———— "Kant and the extragalactic nebulae." *Quarterly Journal of the Royal Astronomical Society* 8 (1967): 48.

———— *The Natural Philosophy of Time.* 2nd ed. Oxford: Clarendon Press, 1980.

Whitrow, G. J., and B. D. Yallop. "The background radiation in homogeneous isotropic world-models." Pt. I. *Monthly Notices of the Royal Astronomical Society* 127 (1964): 301. Pt. II, 130 (1965): 31.

Whittaker, E. T. *A History of Ether and Electricity.* Vol. 1, *The Classical Theories.* Vol. 2, *The Modern Theories 1900–1926.* London: Nelson, 1951.

———— *From Euclid to Eddington.* New York: Dover Publications, 1958.

Whyte, L. L., A. G. Wilson, and D. Wilson, eds. *Hierarchical Structures.* New York: Elsevier, 1969.

Williams. L. P. *Michael Faraday, a Biography.* New York: Basic Books, 1965.

——— "Michael Faraday and the physics of 100 years ago." *Science* 156 (1967): 1335.

Wilson, C. A. *William Heytesbury.* Madison: University of Wisconsin Press, 1956.

——— "How did Kepler discover his first two laws?" *Scientific American* 226 (March 1972): 92.

Wolf, A. *A History of Science, Technology, and Philosophy in the Sixteenth and Seventeenth Centuries.* London: Allen and Unwin, 1935.

——— *A History of Science, Technology, and Philosophy in the XVIIIth Century.* London: Allen and Unwin, 1938.

Woodcroft, B., ed. *The Pneumatics of Hero of Alexandria.* Introduction M. B. Hall. London: Macdonald, 1971.

Wren, C. "The life of Sir Christopher Wren." In *Parentalia: Or, Memoirs of the Family of the Wrens.* London: Stephen Wren, 1750. Reprinted, Farnborough: Gregg Press, 1965.

Wright, T. *An Original Theory or New Hypothesis of the Universe, 1750.* Introduction M. A. Hoskin. New York: Elsevier, 1971.

Yates, F. A. *Giordano Bruno and the Hermetic Tradition.* Chicago: University of Chicago Press, 1964.

Young, T. *Miscellaneous Works.* Ed. G. Peacock. London: John Murray, 1855.

Zeller, E. *Outlines of the History of Greek Philosophy.* Trans. L. R. Palmer. Rev. W. Nestle. New York: Dover Publications, 1980.

Zöllner, J. C. F. *Über die Natur der Cometen.* Leipzig: Staackmann, 1883.

Zwicky, F. "On the red shift of spectral lines through interstellar space." *Proceedings of the National Academy of Sciences* 15 (1929): 773.

Index